Minerals and Rocks 20

Editor in Chief
P. J. Wyllie, Pasadena, CA

Editors
A. El Goresy, Heidelberg
W. von Engelhardt, Tübingen · T. Hahn, Aachen

Donald J. DePaolo

Neodymium Isotope Geochemistry
An Introduction

With 104 Figures

Springer-Verlag
Berlin Heidelberg New York
London Paris Tokyo

Professor Dr. DONALD J. DEPAOLO
Department of Geology and Geophysics
University of California
Berkeley, CA 94720, USA

Volumes 1 to 9 in this series appeared under the title
Minerals, Rocks and Inorganic Materials

ISBN 3-540-18648-4 Springer-Verlag Berlin Heidelberg New York
ISBN 0-387-18648-4 Springer-Verlag New York Berlin Heidelberg

Library of Congress Cataloging in Publication Data. DePaolo, Donald J., 1951–
Neodymium isotope geochemistry: an introduction / Donald J. DePaolo. p. cm. –
(Minerals and rocks; 20) Bibliography: p. Includes index. ISBN 0-387-18648-4
(U.S.) 1. Geochemistry. 2. Isotope geology. 3. Neodymium. I. Title. II. Series.
QE515.D47 1988   551.9 – dc19   88-12328

Typesetting: K + V Fotosatz GmbH, Beerfelden
Offsetprinting and bookbinding: Brühlsche Universitätsdruckerei, Giessen
2123/3130-543210

To Lynn,
Tara, and Daniel

Preface

This monograph was begun with two objectives in mind. The first was to provide a review of research involving the application of neodymium isotopic measurements to problems in earth science. In the process of organizing to do this, I realized that the research in this field had produced a need for an updated review of the underlying paradigms. This need had arisen because of the special properties of the samarium-neodymium isotopic system, and because the research had transgressed the traditional boundaries between the subfields of earth science. Without such a review, the significance of the results seemed likely to remain unnecessarily obscure to interested scientists from related disciplines. Consequently, the second objective became the provision of a theoretical framework for the application of neodymium isotopic studies. Much of what this contains is not new, but it is drawn together here for the first time.

At the time the writing was initiated, the literature of the field was still relatively limited. Over the past 5 years it has grown enormously. Considering the rate at which the writing progressed, it became clear that this could not be a fully up-to-date review and still reach completion. The selection of material for the review sections is biased toward earlier studies.

Part I presents most of the background information. Chapter 1 provides some historical context, a qualitative description of the niche filled by Nd isotopic studies, and some basic properties of Sm and Nd. Chapters 2 through 4 contain the "nuts and bolts." Particularly important aspects are the relationship between the notation and models of planetary evolution (Chap. 3) and the relationship between models for igneous processes (Chap. 4) and models for isotopic evolution. Part II contains an overview of the results as they apply to problems on a planetary scale, including applications to sedimentary rocks and oceanography. Part III contains discussion of results related to problems of the origins of igneous rocks.

The manuscript was written while I was at the Department of Earth and Space Sciences at the University of California, Los Angeles. Harriet Arnoff helped with the typing of early versions. Jean Sells did most of the drafting, with contributions from Vickie Doyle-Jones. Terry Gautesen-Borg transferred the text onto personal computer discs, without which it never would have been completed. Lynn Ingram-DePaolo helped with proofreading. G. Lang Farmer, Bruce Nelson, Vickie Bennett, Brian Stewart, Frank Perry, Don Musselwhite, and Steve Semken contributed data and ideas that are integrated into the text. Research support from the U.S. National Science Foundation was instrumental in allowing this monograph to be prepared.

Berkeley, California   DONALD J. DEPAOLO
May 1988

Contents

Part I   Principles and Processes

Chapter 1   Introduction

1.1 Geologic Perspective

The elements samarium and neodymium have little economic value and are present in only small amounts in most rocks and minerals, yet because of their nuclear and chemical properties, they provide access to information about the origin and evolution of the earth, the moon, and the solar system. They can be considered, like the microscope or the seismograph, to be a tool that can be applied to problems in geology and planetary science. As with other instruments of science, the interpretations of the natural variations of samarium and neodymium isotopic and chemical abundances are governed by principles, models, and assumptions. These and some of the results of the studies to which they have been applied are the subjects of this book.

The geologic utility of these elements stems from the fact that one isotope of samarium (^{147}Sm) is radioactive, and decays by the a-process to one isotope of neodymium (^{143}Nd). The decay proceeds at a very slow rate, the half-life of ^{147}Sm (106 billion years) being some 23 times greater than the age of the earth. Over geologic time, the decay of ^{147}Sm to ^{143}Nd results in changes in the abundance of ^{143}Nd relative to the other Nd isotopes. These small variations can be precisely measured with modern mass spectrometers. The regularity of radioactive decay makes the ^{147}Sm-^{143}Nd pair a timepiece, and one of its most important applications is the determination of the ages of rocks, or more generally, the determination of the timing of major events in the chemical evolution of planets. Sm-Nd studies of igneous rocks can also be used as probes into the earth's interior, capable of sensing the existence of layers or domains of different chemical composition as well as tracing the origin and development of these structures back through time. It is this capability, of determining not only the existence of structure in the earth's interior, but also of determining its age and origin, that makes isotopic studies of ^{147}Sm-^{143}Nd and other similar chronometric nuclide pairs valuable to planetary sciences. Like fossils, these nuclides preserve in the rock record information that allows the ancestry of the modern solid earth to be investigated. The results of these studies relate directly to such diverse fields as igneous, metamorphic and sedimentary petrology and geochemistry, paleogeography, oceanography, crustal tectonics, and mantle geophysics.

1.2 Phylogeny

The application of naturally-occurring radioactive elements and their decay products to geology can be traced back to the discovery of radioactivity in the late 19th century by Becquerel and the Curies. A concise and readable account of this early history can be found in Faure (1977). Isotope geology during the first half of the 20th century was based mainly on the U-Pb method (Rankama 1954, 1963; Russell and Farquhar 1960). Perhaps the biggest single development during this time was the design and construction by Nier (1940, 1947) of mass spectrometers to routinely measure isotopic abundances. This led to a flowering of isotopic research in geology during the 1950s and 1960s (Faure 1977), and in particular, to wider application of the Rb-Sr and K-Ar techniques (e.g. Gast 1960; Faure and Hurley 1963; Hedge 1966; Wasserburg 1966; Aldrich and Nier 1948; Wasserburg and Hayden 1955; Wetherill et al. 1955; Hart 1964; McDougall 1966; Goldich 1968; Doe 1970; Faure and Powell 1972). Modern reviews can be found in Faure (1977) as well as O'Nions et al. (1979a), Hart and Brooks (1980), DePaolo (1983c), Zindler and Hart (1986), and Carlson (1987). The questions of concern in this section are: What determines which isotopes are useful in geology, and why was the Sm-Nd method applied only relatively recently?

In general, the number of chronometric nuclide pairs available for use in geologic studies has been limited partly by the small number of nuclides with suitable half-lives, and partly by analytical difficulties. In order for measurements of the isotopic composition of an element in rocks and other materials of geologic interest to be feasible, certain requirements must be met. One requirement is to have chemical methods that can be used to separate the element from rocks in sufficiently pure form. The element must also have properties that will facilitate mass spectrometric analysis. With regard to the latter, the basic requirement for analyzability is that the element become sufficiently ionized in the mass spectrometer to produce a measureable ion beam. Standard methods of thermal ionization (Dempster 1918; Inghram and Chupka 1953; Wilson and Daly 1963) have proven most successful for elements that have low ionization potentials and are significantly less refractory than the standard filament materials, tantalum, rhenium, tungsten, and platinum. In some cases special techniques have been developed to ionize nonrefractory elements with large ionization potentials (Crouch 1963; Inghram and Chupka 1953). The most notable example is the use of a silica gel medium to enhance the thermal ionization of Pb (Cameron et al. 1969). However, understanding of the ionization process is limited, and special techniques are usually developed simply by trial and error. Satisfactory methods for the ionization of refractory elements with high ionization potentials are still lacking. Examples are osmium and hafnium, which are of considerable geochemical interest, and have become usable only recently (Allegre et al. 1980; Patchett and Tatsumoto 1980a, b) due to technical limitations. Some of the properties of the elements used in geochronology are summarized in Table 1.1, together

Table 1.1. Chemical properties relevant to mass spectrometry

	Melting point (°C)	First I.P. (eV)	Work function (eV)	Max. ionization efficiency (%)[a]	Smallest analyzable sample (ng)[b]	Crustal abundance (ppm)
Filament materials						
Re	3175	7.88	4.96	–	–	0.001
Pt	1772	9.0	5.65	–	–	0.01
Ta	2996	7.89	4.25	–	–	2
W	3410	7.98	4.55	–	–	1
Daughter elements						
Pb	327.5	7.42	4.25	2 (Si gel)	<1	7
Sr	772	5.69	2.59	0.5	200	400
Nd[c]	1016	5.49	3.3	5 (oxide)	20	20
Hf	2220	7.0	3.9	0.01	1000	3
Os	3015	8.7	4.82	<0.01?	?	0.005
Ca	842	6.11	2.87	0.1	5000	5×10^4
Parent elements						
U[c]	1132	6.08	3.63	0.1	<1	1
Th	1800	6.95	3.4	<0.1	<1	4
Rb	38.9	4.18	2.16	~100	<1	35
Sm[c]	1073	5.63	3.2	1	<1	4
Lu	1663	5.43	3.3	0.1	<1	0.4
K	63.6	4.34	2.30	~100	<1	10^4

[a] Maximum ionization efficiency is here defined as the ratio of the total number of "measurable ions" reaching the mass spectrometer ion collector to the number of atoms initially present on the source filament. "Measurable ions" refers to that portion of the run when the ion beam is of sufficient intensity and stability to yield precise isotopic ratio data.

[b] Measurements can be made on smaller samples of the parent element because only the concentration need be determined. Precise isotropic ratio measurements on the daughter element generally require larger samples.

[c] Elements that can be measured as oxide species (e.g. NdO^+, UO_2^+). The ionization energy can be more favorable for the oxide, although the actual values are not available.

with an indication of the ease with which they can be analyzed by thermal ionization mass spectrometry.

Geochemical considerations are closely tied to the technical problems. The variability of the parent/daughter ratio in nature, which is a function of the chemical contrast between the two elements, and the half-life, determines the magnitude of isotopic variations that can occur. The variations must be substantially larger than the precision of a measurement in order to be useful. The attainable precision is dictated by the properties of the element as discussed above, combined with limitations imposed by the design and construction of the mass spectrometer. Finally, the natural abundances of the elements are important, since elements present in extremely low concentrations in rocks can present problems in chemical separation procedures.

The relatively recent application of the Sm-Nd method to problems in geology was made possible by a series of technical developments which took place over 3 decades. As is true of many techniques in the earth sciences, the lead-up developments originated in other fields, chemistry and physics in this case, and the implementation proceeded in small steps, over a substantial time period, as those developments were made available to geochemists who could identify the potential applications.

The first step toward the eventual use of Sm-Nd studies in geology can be traced to the development of ion exchange methods for the efficient separation of rare earth elements (REE). These were first developed in the 1940s to separate uranium fission products produced in nuclear reactors (Spedding et al. 1947; see reviews by Peppard 1961 and Powell 1961). In the 1960s these procedures were employed to partially separate REE for the purpose of determining their abundances in rocks by neutron activation (Haskin et al. 1966). The methods for high purity separation of individual rare earths for mass spectrometric measurements were first applied by Eugster et al. (1970) who studied neutron-induced isotopic effects in Gd and Sm from meteorites and lunar samples.

The second step involved the improvement of the precision of isotopic abundance ratio measurements. The variation of the abundance of ^{143}Nd in nature could be estimated to be small, of the order of a few tenths of a percent, due to the long half-life of ^{147}Sm and the limited variation of Sm/Nd. The usefulness of the Sm-Nd system depended upon the ability to make precise measurements of Nd isotopic abundances. The typical precision for isotopic ratio measurements that was attainable with the thermal ionization mass spectrometers used until the 1960s was $\pm 0.1\%$ (e.g. Hedge 1966); about the same magnitude as the variations in ^{143}Nd. This particular limitation was imposed by the stability of the ion beam, the detector characteristics, and the data acquisition procedures rather than by the properties of the elements. Measurements were made by manually reading peak heights from a chart recorder while the mass spectrum was scanned in analog mode by varying the magnetic field. The development of a computer-interfaced mass spectrometer with rapid magnetic field switching and digital data acquisition by Wasserburg et al. (1969) improved the precision by a factor of 30 and brought the use of the Sm-Nd system into the range of feasibility.

In pioneering studies of the meteorite Juvinas and a lunar basalt, Lugmair (1974) and Lugmair et al. (1975a) combined the chemical methods of Eugster et al. (1970) with modern mass spectrometric techniques to become the first to exploit the Sm-Nd system in geochemistry. Their studies demonstrated that the system could be used to obtain precise ages of old rocks, and also yielded important evidence in support of the early chemical differentiation of the moon.

As shown in Table 1.1, the chemical properties of Nd are favorable for precise isotopic measurements, especially the low ionization potential and relatively low melting point. In fact, using standard techniques, Nd samples can be analyzed with such ease that measurements can be made to the precision limits of present-day mass spectrometers on only 2×10^{-8} g of separated Nd.

For a typical rock sample that has a Nd concentration of 20 parts per million (ppm) by weight, this amount of Nd is supplied by only 1 mg of rock; a piece the size of the ball on a ball-point pen.

1.3 Characteristics of the Sm-Nd System and Comparison with Rb-Sr and U-Th-Pb

The application of Sm-Nd isotopes draws on an extensive groundwork laid by the previous studies of ^{87}Rb-^{86}Sr-, ^{235}U-^{207}Pb, ^{238}U-^{206}Pb, and ^{232}Th-^{208}Pb. The value of ^{147}Sm-^{143}Nd relative to these other pairs can be likened to the value of an electron microscope relative to an optical microscope or an X-ray diffractometer. Just as each instrument is sensitive to a certain aspect of mineral structure which cannot be detected with the other instruments, ^{147}Sm-^{143}Nd is sensitive to certain geologic processes and insensitive to others, in a manner that is complementary to the other nuclide pairs. The principles are the same, the differences lie only in the chemical properties of Sm and Nd, which enable them to magnify the effects of some processes which would be unresolvable using the other nuclide pairs, while being oblivious to other processes that are better investigated with the other nuclides.

The study of Nd isotopic variations is unique among isotopic investigations in that the first measurements were preceded by more than a decade of intensive study of the distribution of Sm and Nd and the other rare-earth elements in rocks and minerals (e.g. Haskin et al. 1966; Schilling and Winchester 1967; Frey et al. 1971; Taylor 1964; Gast 1968; Kay and Gast 1973; see Haskin and Paster 1979; Hanson 1980; and Frey 1982 for reviews). REE studies have contributed immensely to knowledge of the petrogenesis of a wide variety of rock types, and had developed into an important geochemical field well before the first Sm-Nd isotopic measurements were made. The Sm-Nd studies came at a time when the geochemistry of both parent and daughter were already well understood, providing the opportunity to add time constraints to an already valuable geochemical technique.

1.3.1 General Geochemistry and Cosmochemistry

For Sm-Nd, both the parent and daughter elements are refractory in the sense that they are thought to occur early in the sequence of elements that would have condensed from the cooling solar nebula 4.5 billion years ago (Grossman and Larimer 1974). Consequently, parent/daughter fractionation during condensation of the solar nebula should have been negligible because both elements were quantitatively transferred into the early-formed solid bodies that later accreted to form the planets. Boynton (1975) has calculated that only under conditions of extreme disequilibrium will Sm and Nd be fractionated during condensation. Therefore, most of the variation of Sm/Nd in the earth today is a result of the earth's internal differentiation rather than a result of condensation processes (Chap. 3). In contrast, Pb is much more volatile than U and Th, and Rb more volatile than Sr under the conditions thought to have prevailed in the solar nebula, so large parent/daughter fractionation may have occurred during condensation.

Available data support these inferences. The ratio of U/Pb in the earth is 25 times higher than the solar value, and Rb/Sr is about 8 times lower. The moon is a more extreme example. The ratio U/Pb in the moon is perhaps 100 times higher, and the ratio Rb/Sr 30 times lower than the solar ratio. By comparison, isotopic data on terrestrial basalts (e.g. Hofmann and Hart 1978) suggest that the variability of Rb/Sr in the earth's mantle today is of the order of a factor of two, and the variability of U/Pb is even less. Since these variations are dwarfed by the possible variations produced during condensation, it is more difficult to distinguish variations due to planet formation processes from those caused by the planet's internal differentiation. On the other hand, the large parent/daughter fractionations in the Rb-Sr and U-Th-Pb systems provide precise information on the time of planet formation, whereas the Sm-Nd clock is insensitive to that event (Chap. 3).

Sm and Nd are lithophile in their geochemical character. In contrast, U and Th are lithophile, but Pb is both lithophile and chalcophile. If planetary core formation involved the incorporation of sulfides into the core (Murthy and Hall 1970), large U-Th-Pb fractionations might be expected, but Sm-Nd fractionation would not be expected. These inferences, of course, are based on the near-surface properties of the elements, which might not apply to their behavior during core formation if high pressures and temperatures were involved (e.g. Lewis 1971).

In summary, Sm/Nd fractionation during planetary evolution probably occurs only as a result of magmatism in the silicate portion of the planets, and not during condensation or core formation. This is different from Rb/Sr,

Table 1.2. Concentrations (ppm) in some typical terrestrial rooks and in the silicate portion of the earth (mantle and crust)

Element	Silicate earth	MORB	Oceanic alk. basalt	Granite (G-1)	Av. Shale
Sm	0.29	3.5	8.5	8.3	5.8
Nd	0.90	10	44	55	24
Rb	0.5	1.1	30	220	140
Sr	17	130	900	250	300
U	0.018	0.10	1	13.4	3.7
Th	0.070	0.20	4	50	12
Pb	0.13	0.5	6	48	20
Lu	0.048	0.45	0.2	0.19	0.6
Hf	0.23	2.3	5	5.2	5

Fig. 1.1. Enrichment factors of elements important in geochronology, relative to the concentrations in average chondritic meteorites. The normalizing chondritic values used are those listed for the silicate earth (Table 1.2) divided by 1.5 except for Rb (2.5 ppm) and Pb (3.8 ppm)

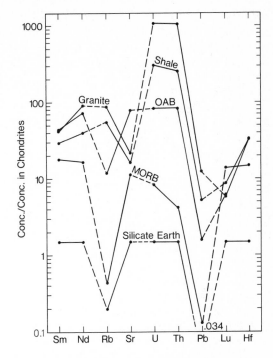

U/Pb, and Th/Pb, each of which can be affected by condensation and/or core formation. The abundances of Sm, Nd, Rb, Sr, U, Th, and Pb are given in Table 1.2 for chondrites, the estimated "bulk silicate earth" and some common rock types. The depletion of the silicate part of the earth in Rb and Pb is noteworthy. However, the Pb abundance, estimated from abundances in surface rocks and mantle samples, could be too low for the whole earth if the earth's core contains a large amount of Pb. Figure 1.1 shows the enrichment factors relative to the average chondritic abundances. Crustal rocks are highly enriched in all of these elements relative to the silicate earth, and with the exception of Pb, relative to chondrites.

1.3.2 Petrogenetic Significance of the Parent/Daughter Ratio

Because the REE differ chemically only with respect to ionic radius, which decreases regularly with increasing atomic number, the fractionation that occurs within the REE group is systematic. Some general features of rare-earth abundance patterns are illustrated in Fig. 1.2. The abundances measured in the rocks are normalized to the average abundances of chondritic meteorites and are plotted against atomic number from La (57) to Lu (71). The abundances of the light REE in rocks vary over several orders of magnitude, whereas the heavy REE show much less variability. In general, the curves are

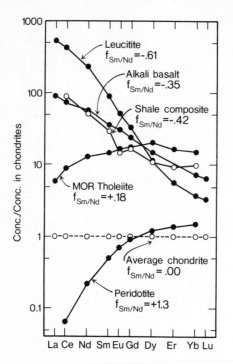

Fig. 1.2. Chondrite-normalized rare-earth element patterns of some terrestrial rock samples. The parameter $f_{Sm/Nd}$ gives the difference of the Sm/Nd ratio relative to chondrites (see Chap. 3). The rare-earth elements are listed by increasing atomic number from *left to right*. Data are from Kay and Gast (1973), Frey et al. (1971), O'Nions et al. (1979a), Haskin and Paster (1979)

smooth functions of atomic number. The pattern for a composite of North American shales, with light REE enriched relative to heavy REE, is typical of the vast majority of continental crustal rocks and is considered to be representative of continental crust (Haskin et al. 1966). The lower curve, a sample of peridotite derived from the upper mantle, has a distinctly different abundance pattern, with light REE relatively depleted, and low abundances of all rare earths. The other three patterns shown in Fig. 1.2 are from basaltic rocks. These examples demonstrate that the abundance patterns are highly variable even among rocks that are classed as "basalt". The REE patterns vary systematically with other aspects of the bulk chemistry.

Both Sm and Nd are light rare earths, and their position within the group is such that the overall abundance pattern is usually accurately reflected by the Sm/Nd ratio. Low Sm/Nd is indicative of a light REE-enriched pattern, and high Sm/Nd is characteristic of a light REE-depleted pattern (relative to heavy REE). This is an important property of Sm and Nd, because it means that information gained from Nd isotopic studies provides information about the overall REE chemistry of the parent reservoirs from which the rocks were derived (Chap. 3). This information can then be integrated with the broad base of information about the behavior of REE in petrogenetic processes to help understand both the petrogenesis of the rocks and the evolution of the parent reservoir (Chap. 4). The link between REE fractionation and igneous petrogenesis is provided by extensive studies of REE partitioning between minerals and magma, and theoretical models of trace element fractionation during magma genesis (Schilling and Winchester 1967; Schnetzler and Philpotts 1970; Gast 1968; Kay and Gast 1973; Weill and McKay 1975; Drake and Weill 1975; Hanson 1980; Allegre and Minster 1978; DePaolo 1981d; Richter 1986).

The Rb-Sr system is similar to Sm-Nd in that the properties of Rb, an alkali element, differ from the properties of Sr, an alkaline earth, in ways that allow Rb/Sr fractionation to be interpreted in terms of petrogenetic processes that tend to separate the Group I and II elements. However, the correlation of K/Ca, Cs/Ba, and Na/Mg fractionation with Rb/Sr fractionation is only approximate at best. The most important consideration is that the behavior of Rb and Sr during magma genesis is now quite well understood as a result of experimental studies of partitioning between minerals and magma (see Irving 1978 for review and Chap. 9). The U/Pb-Th/Pb system is quite different in this regard. The geochemistry of U, Th, and Pb has been studied extensively but there are not close chemical homologues of these elements that can be used to infer the broader petrogenetic character of the fractionation processes. Furthermore, accurate determinations of partition coefficients for minerals and magma have only recently been undertaken (e.g. Benjamin et al. 1980). On the other hand, the U-Pb system has the unique advantage of a coupled pair of uranium isotopes, both of which decay to Pb isotopes. The systematics can often be used to unravel complicated geologic histories (e.g. Doe 1970; Tera and Wasserburg 1974).

1.3.3 Magnitude of Isotopic Variations

The relative chemical similarity of Sm and Nd results in rather modest separation in nature, which in turn yields only small isotopic effects. The typical range of the weight ratio of Sm/Nd in rocks is from about 0.1 to 0.5 (cf. Richard et al. 1976). Minerals generally show about the same variation, although some rare minerals, such as xenotime, may have Sm/Nd somewhat greater than one. By contrast, in the U-Th-Pb system, the strong chemical differences between the parent and daughter elements result in the very large fractionation during geologic processes. The large fractionation causes enormous variations in the isotopic composition of Pb, which permit a clear determination of the times of fractionation (e.g. Krogh 1982a, b). Rb and Sr are similar to U-Th-Pb in that they also have extremely different chemical properties and can be greatly fractionated during magmatic differentiation, metamorphism, and weathering. Rb/Sr in rocks and minerals can vary from nearly zero to 10^4 or more. Consequently, isotopic variations in Sr can be thousands of times larger than the largest possible isotopic variations in Nd.

1.3.4 Behavior in the Near-Surface Environment

An advantage of the Sm-Nd system relative to Rb-Sr and U-Th-Pb is that the rare earths are less mobile under near-surface conditions. Therefore, the systematics are less likely to be disturbed. The "mobility" of an element is a loosely defined property that depends on several factors. A particular problem with the U-Th-Pb system is that the U and the radiogenic Pb are typically concentrated in minor minerals. In contrast, the bulk of the Sm and Nd is more often contained in the major minerals of rocks. Smaller grains are generally more susceptible to leaching or diffusive loss than are the larger mineral grains of the major phases. Furthermore, the large amount of energy released by U decay tends to cause substantial damage to the crystal structure and therefore enhances diffusive loss of U and Pb. The energy released by U decay in the lifetime of a 2.7-billion-year-old rock is about 1.8×10^6 kcal g^{-1} U or about 2000 times more than is released by ^{147}Sm decay in the same period. Since the U-bearing minor minerals can have U concentrations that are much higher than the Sm concentrations in the major minerals, the factor per unit volume can be much higher. Severe Pb loss from zircons associated with metamictization (radiation damage) has been amply demonstrated (e.g. Silver and Deutsch 1963). It is the coupled pair of radioactive U isotopes and the apparently systematic way in which Pb is lost from zircons that make them useful in geochronology.

An additional factor that contributes to mobility is the chemical incongruence of parent and daughter. Radiogenic Pb atoms must occupy structural sites in minerals that may have been chemically favorable for U atoms, but are unsuitable for the Pb atoms. These Pb atoms are probably not bound tightly in the mineral structure and therefore may be especially mobile. Furthermore, the decay of U atoms to Pb involves a large number of intermediate daughter products, some of which are gases that are particularly mobile. Similar problems exist for Rb and Sr. Although Rb decay releases relatively little energy, the extremely different chemical properties result in radiogenic ^{87}Sr atoms being located in unfavorable crystal sites. In the case of phyllosilicates, which are important in Rb-Sr dating, the Rb atoms and hence also the radiogenic ^{87}Sr atoms tend to occupy crystal sites that are easily leachable, and in some cases even exchangeable. The subtle chemical differences between Sm and Nd are probably insignificant in comparison, so a Nd daughter atom can occupy the same site as its Sm parent with little energy change. However, the radiogenic ^{143}Nd nucleus does experience a substantial recoil from the 2.23 MeV α-decay, whereas ^{87}Sr would experience essentially no recoil from the β-decay.

Mobility of atoms may also be a function of their solubility in groundwater and hydrothermal fluids that circulate through rocks. Only a small amount of data exist on this problem for REE, and there is presently only a sketchy understanding of the mechanisms of trace element redistribution during metamorphism and weathering (e.g. Fletcher and Hofmann 1974; Collerson and Fryer 1978; Clauer et al. 1982). The relative solubilities of rare earths versus Rb, Sr, U, Th, and Pb are poorly known and should be a function of fluid composition and other conditions. Redistribution of Rb, Sr, U, Th, and Pb in progressive metamorphism has been documented (e.g. Lanphere et al. 1963; Hart 1964; Doe 1970); however, a detailed understanding of the processes involved and knowledge of the relative mobility of Sm and Nd during metamorphism vis-à-vis these other elements is still lacking. It is known that the concentration of rare earths, as well as Th and Pb, in surface waters is normally lower by many orders of magnitude than that of U, Rb, and Sr (Mason and Moore 1982).

1.4 Isotopic Abundances of Sm and Nd

1.4.1 Summary of Recent Measurements

Geochemical studies of isotopic variations in Sm and Nd in terrestrial, lunar, and meteoritic materials have provided precise values for the relative abundances of the isotopes of these elements. The isotopic composition of Sm, measured by Russ et al. (1971) and Russ (1974), and the isotopic composition of Nd, measured by DePaolo and Wasserburg (1976a, b), are given in Table 1.3 (line A for Nd) expressed as isotopic abundance ratios (or equivalently, "isotopic ratios") relative to a selected normalization isotope that is neither radioactive nor radiogenic, ^{144}Nd for neodymium and ^{154}Sm for samarium. Other determinations of Nd isotopic abundances based on multiple measurements have been reported by Nakamura et al. (1976) and Lugmair et al. (1976).

Figure 1.3, adapted from the Chart of the Nuclides (General Electric Company 1977), shows the isotopes of Sm and Nd and gives the number of protons (Z) and the number of neutrons (N) for each nuclide as well as approximate abundances. There are two points of interest here. First, note that not only ^{147}Sm, but also ^{148}Sm, ^{149}Sm, and ^{144}Nd are listed as being radioactive. These latter three nuclides α-decay to ^{144}Nd, ^{145}Nd, and ^{140}Ce, respectively. It might be expected that these additional decays must be accounted for, especially those affecting ^{144}Nd, since the ratio ^{143}Nd/^{144}Nd is the one that is of geochronological interest. However, the half-lives are so long that all three can be safely treated as being stable nuclides. For example, over the entire 4.5-billion-year history of the earth, the abundance of ^{144}Nd has decreased by only about 0.00015% due to radioactive decay. This amount is about 20 times smaller than the current analytical precision of measurements of the ^{143}Nd/^{144}Nd ratio. Also of note are the thermal neutron capture cross-sections (σ) of the various nuclides. The cosmic-ray produced neutron flux in the surficial layers of the lunar regolith is large enough so that the isotopic abundances of Sm and Nd can be affected over the long times available for exposure on the lunar surface (Russ et al. 1971; Lugmair et al. 1975a). The

Table 1.3. Isotopic ratios and isotopic abundances for Nd and Sm[a]

	^{142}Nd/^{144}Nd	^{143}Nd/^{144}Nd	^{145}Nd/^{144}Nd	^{146}Nd/^{144}Nd	^{148}Nd/^{144}Nd	^{150}Nd/^{144}Nd	Atomic weight	Ab(^{143}Nd)[b]	Ab(^{144}Nd)[c]
A.	1.138266 ± 6	0.511836	0.348964 ± 5	0.724110 ± 5	0.243080 ± 4	0.238581	144.256	0.12172	0.23782
B.	1.141830	0.512636	0.348425	0.721857	0.241572	0.236368	144.240	0.12198	0.23794
	(1.141722)		(0.348475)	(0.721906)		(0.236436)			
A'.	1.138329	0.511852	0.348954	0.72411	0.243064	0.238594	144.256	0.12172	0.23782
	(1.138332)		(0.348945)	(0.724127)	(0.243064)				
B'.	1.141854	0.512644	0.348416	0.721882	0.241572	0.236404	144.240	0.12198	0.23794

	^{144}Sm/^{154}Sm	^{147}Sm/^{154}Sm	^{148}Sm/^{154}Sm	^{149}Sm/^{154}Sm	^{150}Sm/^{154}Sm	^{152}Sm/^{154}Sm	Atomic weight	Ab(^{147}Sm)
C.	0.13516 ± 1	0.65918	0.49419 ± 2	0.60750 ± 2	0.32440 ± 2	1.17537 ± 4	150.35	0.14995
	(0.13520)		(0.49421)	(0.60751)	(0.32440)	(1.17532)		

[a] A = DePaolo and Wasserburg (1976a); normalized to ^{148}Nd/^{144}Nd = 0.241572. Numbers in parentheses are those reported by Carlson et al. (1978).
B = Ratios from line A recorrected using ^{18}O/^{16}O = 0.002045, ^{17}O/^{16}O = 0.000371.
A' = Ratios from line A recorrected using ^{18}O/^{16}O = 0.00210, ^{17}O/^{16}O = 0.000384. Numbers in parentheses are those reported by DePaolo (1981b).
B' = Ratios from line A' renormalized to ^{148}Nd/^{144}Nd = 0.241572, which is approximately equivalent to ^{146}Nd/^{144}Nd = 0.72190.
C = Russ (1974); numbers in parentheses measured by DePaolo (1981b) and Lugmair et al. (1975a).
[b] Varies in nature.
[c] Varies in nature, but only a negligible amount (±0.01%).

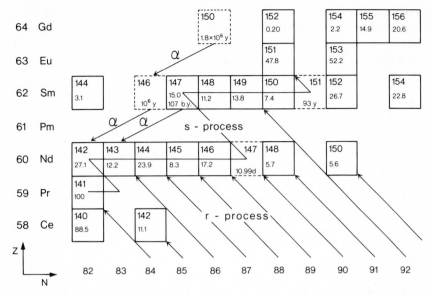

Fig. 1.3. Chart of the number of protons (Z), the number of neutrons (N), the approximate abundances (in percent), and radioactive decays for the nuclides in the mass region of Sm and Nd. The nucleosynthetic pathways (r, s) (see e.g. Clayton 1977) affecting the Sm and Nd isotopes are also shown (After O'Nions et al. 1979)

abundance of ^{149}Sm has been found to be as much as 1% smaller than normal in lunar soil samples (Lunatic Asylum 1978) and has been used along with ^{155}Gd and ^{157}Gd in lunar stratigraphic studies (Russ 1974). Neutron capture affects ^{143}Nd by an amount that is much smaller, but significant in comparison to the analytical uncertainty in ^{143}Nd/^{144}Nd.

With the exception of the few localities where natural uranium reactors have been found (Naudet 1978), terrestrial materials exhibit no variation of either Sm or Nd isotopic composition except for shifts in the ^{143}Nd abundance due to ^{147}Sm decay. Lunar and meteoritic materials, however, show shifts in both Sm and Nd isotopic composition due to neutron capture. In addition, distinctive variations have been found in some meteoritic materials due to the preservation of incompletely mixed materials from different nucleosynthetic sources (McCulloch and Wasserburg 1978a, b; Lugmair et al. 1978) and (possibly) from the decay of short-lived ^{146}Sm to ^{142}Nd (Lugmair et al. 1975b; Lugmair and Marti 1977; Jacobsen and Wasserburg 1984).

1.4.2 Mass Discrimination Corrections and Interlaboratory Data Comparison

Mass spectrometric determination of isotopic abundance is affected by isotopic fractionation in the mass spectrometer, the magnitude of which can vary from analysis to analysis and even during a single analysis. Because of this, precise, reproducible values for isotopic abundance ratios can be obtained only if the measured ratios are corrected to an assigned value of the abundance ratio of two stable nonradiogenic isotopes (cf. Faure 1977). For example, the data for Nd in Table 1.3 (lines A and A') were adjusted to give a value of $^{150}Nd/^{142}Nd = 0.209600$. This value is the average raw $^{150}Nd/^{142}Nd$ measured over a large number of runs at the California Institute of Technology (DePaolo and Wasserburg 1976a). Whether this corresponds to the actual value in nature is not known. In other laboratories, Nd isotopic abundance ratios are normalized to $^{148}Nd/^{144}Nd = 0.241572$ (Lugmair et al. 1975a), $^{146}Nd/^{144}Nd = 0.72190$ (O'Nions et al. 1977) or $^{146}Nd/^{142}Nd = 0.63613$ (DePaolo 1981b). Since Nd has six nonradiogenic isotopes, other pairs of isotopes can be chosen for normalization. $^{150}Nd/^{142}Nd$ was chosen for normalization by DePaolo and Wasserburg (1976a) because the difference of eight mass units between the two isotopes makes $^{150}Nd/^{142}Nd$ a sensitive indicator of the fractionation per mass unit. A precise estimate of the fractionation per mass unit is important, since the precision obtainable on measurements of $^{143}Nd/^{144}Nd$ is more than ten times smaller than the variations caused by fractionation in the mass spectrometer during a single analysis.

In order to compare measurements reported in the literature from different laboratories, the differences in normalization procedures must be taken into account. A change in the normalization value for any element changes the calculated relative proportions of the isotopes, and therefore changes the atomic weight of the element. Assuming a linear relationship, the ratio calculated for a pair of isotopes, i and j, using one normalization (normalization A) can be converted to the equivalent ratio in a second normalization (normalization B) by the transformation:

$$(R_{ij})_B = (R_{ijA}) [1 + a_{AB}(i - j)]^{-1} ,$$

where R_{ij} is the abundance ratio of isotopes with mass numbers i and j. By substituting values of R_{ij} for the two normalizations into this equation, a can be calculated. This value can then be used to correct all of the other isotopic ratios. To determine a as accurately as possible, a ratio R_{ij} should be selected for which the difference $i - j$ is large. As an example of the application of this equation, Lugmair et al. (1975a) normalized their data to $^{148}Nd/^{144}Nd = 0.241572$, whereas DePaolo and Wasserburg (1976a, b) obtained an average value of $^{148}Nd/^{144}Nd = 0.243080$ in their normalization. Substituting into the above equation:

$$(^{148}Nd/^{144}Nd)_L = (^{148}Nd/^{144}Nd)_{DW} [1 + a_{L-DW}(148 - 144)]^{-1}$$

gives $a_{L-DW} = 0.001561$. The $^{143}Nd/^{144}Nd$ in the two normalizations are therefore related by:

$$(^{143}Nd/^{144}Nd)_L = (^{143}Nd/^{144}Nd)_{DW} [1 + 0.001561 (143 - 144)]^{-1}$$

$$= (^{143}Nd/^{144}Nd)_{DW} \cdot 1.001563 .$$

All laboratories use one of these two normalizations.

The ratios measured by DePaolo and Wasserburg (1976a) are given in Table 1.3 (line A). The same set of ratios renormalized according to the procedure of Lugmair et al. (1975a) is given in line B. Shown in parentheses in line B are the ratios reported by Carlson et al. (1978). They do not quite agree with the DePaolo and Wasserburg data to the level of quoted precision. The ^{145}Nd/^{144}Nd ratio, for instance, differs by 0.014%, whereas the quoted precision is 0.005%. This gives some indication of the comparability of data from different laboratories (see Jacobsen and Wasserburg 1980a and Chap. 5).

In some laboratories, Nd isotope ratios are measured in the mass spectrometer on the species NdO^+. When this is done, the ratios $^iNd/^jNd$ must be calculated from the measured $^iNdO/^jNdO$ ratios by a procedure that accounts for the isotopic composition of oxygen in the NdO^+ ion beam. This correction requires that the O isotope composition be precisely known. As discussed by Papanastassiou et al. (1977) and in more detail by DePaolo (1978a), the O isotope composition of an NdO^+ ion beam can be somewhat variable, and significantly different from the composition given by Nier (1950). The observed variations are such that there would be no inconsistencies in data collected in one laboratory, but they could affect comparisons made between laboratories measuring NdO^+ and laboratories measuring Nd^+ directly. Differences of up to 0.005% ^{143}Nd/^{144}Nd (equal to the typical analytical uncertainty) could result, depending on the method of normalization. The Nd^+ ion, which is measured in some laboratories, would be the preferred species because it eliminates the oxygen isotope correction. However, the ionization efficiency of Nd^+ is significantly less than that for NdO^+, requiring larger amounts of Nd to obtain precise measurements.

Lines A' and B' of Table 1.3 give the Nd isotope abundance ratios that should be regarded as the best estimates. Line A' gives the ratios from line A recorrected with appropriate oxygen isotope ratios (Wasserburg et al. 1981). Shown in parentheses are ratios measured on Nd^+ by DePaolo (1981b) normalized to ^{146}Nd/^{142}Nd = 0.636130, which are in excellent agreement. Line B' gives the ratios in the Lugmair et al. (1975a) normalization. Normalization using the ratio ^{146}Nd/^{144}Nd = 0.72190 (e.g. O'Nions et al. 1977) can be seen to be essentially identical to the normalization given on line B'. The ratios in Table 1.3 (line A') have been measured on three different mass spectrometers, which provides confidence in their internal consistency.

An additional minor problem is the validity of the mass discrimination procedure discussed above, which incorporates the assumption that the discrimination factor a is mass-independent. Experiments by Russell et al. (1977) have shown that for Ca, a is a function of mass. Certain Nd isotopic effects discussed by DePaolo (1978a) and Wasserburg et al. (1981) are apparently caused by mass-dependent fractionation that is significant at the high precision levels required in Nd isotopic studies. The magnitude of these effects can be of order 0.01% in ^{143}Nd/^{144}Nd.

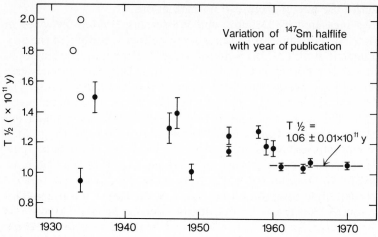

Fig. 1.4. Summary of determinations of the half-life of ^{147}Sm (After Lugmair et al. 1976)

1.5 Decay Constant of ^{147}Sm

The half-life of ^{147}Sm has been measured several times by α-counting techniques. The results are summarized in Fig. 1.4. The most recent determination
by Gupta and McFarlane (1970) yielded a result of 106 ± 2 b.y., which is also
the mean of the most precise determinations (cf. Lugmair and Marti 1978). In
some cases a significant proportion of the total uncertainty in these determinations was the result of uncertainly in the isotopic composition of Sm, which
has since been measured more precisely (Russ et al. 1971). With the incorporation of the more precise isotopic abundances, it appears that the single most
precise determination of the ^{147}Sm half-life is by Wright et al. (1961) and is
105.3 ± 0.3 b.y. However, in accordance with the initial suggestion by Lugmair
and Marti (1978), the mean of 106 ± 1 b.y. ($\lambda = 0.00654$ b.y.$^{-1}$) calculated from
the most recent determinations, is used here. A more precise determination of
the ^{147}Sm half-life using the best available modern techniques would be
desirable, especially since the time resolution with the Sm-Nd method may in
some instances be less than the uncertainty in the half-life (Papanastassiou et
al. 1977; DePaolo and Wasserburg 1979a; and Chap. 2). Uncertainties in the
half-life are particularly important when comparing ages determined with different parent-daughter systems.

Chapter 2 Sm-Nd Dating

2.1 Theory and Requirements

The Sm-Nd system is useful for determining crystallization ages of igneous and metamorphic rocks. It has proven especially valuable for dating mafic and ultramafic rocks that are, in general, difficult to date using other methods. The principles of Sm-Nd dating are identical to those of Rb-Sr dating, which have been discussed in detail elsewhere (e.g. Lanphere et al. 1963; Faure 1977). The underlying assumption is that at some time T_x in the past, all the minerals in a rock have identical values of the ratio $^{143}Nd/^{144}Nd$, but have different ratios of $^{147}Sm/^{144}Nd$. This starting point is shown in Fig. 2.1 by the open circles representing minerals A, B, and C. The open circle labelled "TR" represents the whole rock, which is just the sum of the minerals, and thus must also have the same $^{143}Nd/^{144}Nd$. As time progresses ^{147}Sm atoms decay to ^{143}Nd atoms, so $^{147}Sm/^{144}Nd$ decreases and $^{143}Nd/^{144}Nd$ increases. The positions of the minerals and rock on this diagram move along the arrows shown, which have a slope of -1. For each mineral, $^{143}Nd/^{144}Nd$ as measured today ($T = 0$) is given by:

$$\frac{^{143}Nd}{^{144}Nd_{(0)}} = \frac{^{143}Nd}{^{144}Nd_{(T_x)}} + \frac{^{147}Sm}{^{144}Nd_{(0)}} (e^{\lambda T_x} - 1) \quad ,$$

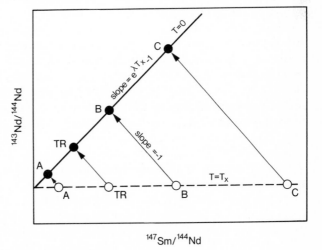

Fig. 2.1. Samarium-neodymium isotopic evolution for the minerals (*A*, *B*, and *C*) and the bulk sample (*TR*) of an igneous rock that crystallized T_x years ago. The time $T = 0$ is the present; $^{143}Nd/^{144}Nd$ ratio that corresponds to $^{147}Sm/^{144}Nd = 0$ is the "initial" $^{143}Nd/^{144}Nd$ ratio of the rock

which is the equation for a straight line of slope $(e^{\lambda T_x} - 1)$ and intercept $^{143}Nd/^{144}Nd(T_x)$. If a mineral had $^{147}Sm/^{144}Nd = 0$ it would simply retain the value $^{143}Nd/^{144}Nd(T_x)$ for all times.

The initial condition is generally expected to hold for minerals of an igneous rock immediately after crystallization. Most magmas should be sufficiently well mixed to ensure that the minerals that crystallize from it will have identical $^{143}Nd/^{144}Nd$, at least on the scale of the sample size needed for analysis (e.g. 10 to 100 cm^3). Chemical fractionation between minerals, however, causes them to have different $^{147}Sm/^{144}Nd$. T_x is referred to as the "crystallization age" of the rock, and $^{143}Nd/^{144}Nd(T_x)$ as the "initial" $^{143}Nd/^{144}Nd$ ratio of the rock. Homogenization of $^{143}Nd/^{144}Nd$ ratios on a scale large compared to the size of typical mineral grains can also occur during high-grade metamorphism.

In order to determine the age of a rock, the slope of the line ABC must be determined precisely. It is therefore necessary to have minerals with as large a difference in $^{147}Sm/^{144}Nd$ as possible. Among common igneous minerals, feldspars normally have relatively low $^{147}Sm/^{144}Nd$, while pyroxenes generally have high $^{147}Sm/^{144}Nd$. Other low-Sm/Nd phases are apatite and monazite, and other high-Sm/Nd phases are garnet, sphene, zircon, and amphiboles. Mafic rocks generally contain abundant plagioclase and pyroxene (or amphibole) and so can usually be dated by Sm-Nd. The typical range of the $^{147}Sm/^{144}Nd$ ratio in the minerals of a rock is about 0.10, although substantially larger ranges have been found. With the current analytical precision of about $\pm 0.003\%$ (e.g. 0.511836 ± 0.000015) the optimum age uncertainty at the 95% confidence level is approximately:

$$\pm \Delta T \approx \pm \lambda_{Sm}^{-1} \frac{\Delta\,^{143}Nd/^{144}Nd}{\Delta\,^{147}Sm/^{144}Nd} = \lambda_{Sm}^{-1} \frac{0.000015}{0.10} \approx \pm 20 \text{ million years} \;.$$

This ± 20 m.y. uncertainty applies regardless of the age of the rock. In terms of relative errors, a 1 b.y.-old-rock could be dated to $\pm 2\%$, whereas a 4 b.y.-old-rock could be dated to $\pm 0.5\%$. In rocks where minerals such as zircon and garnet are present, which can have particularly high $^{147}Sm/^{144}Nd$, the range of $^{147}Sm/^{144}Nd$ could be increased up to 0.5. In that case the age resolution could be ± 4 m.y. (cf. Zindler et al. 1983). If the precision of the measurements of $^{143}Nd/^{144}Nd$ could be improved by a factor of five, which seems likely, the dating capabilities could be improved proportionally. The Sm-Nd system can be used for dating rocks with ages ranging from Cenozoic to earliest Precambrian.

The initial $^{143}Nd/^{144}Nd$ is also an important geochemical parameter. Assuming isotopic equilibrium between magma and solid residue at the time the magma is extracted and emplaced into the crust, the initial $^{143}Nd/^{144}Nd$ of the resultant igneous rock will be identical to that of the source rock at the time of melting. The evolution of $^{143}Nd/^{144}Nd$ in magma *sources* through time can therefore be studied by determining initial ratios for rocks of different ages (Chap. 3). For Sm/Nd it is difficult to obtain precise initial ratios because no common minerals have ^{147}Sm/^{144}Nd near zero. There is usually a long extrapolation to determine the intercept, which results in relatively large uncertainties.

Whole rock Sm-Nd isochrons can also be measured on suites of igneous rocks that have a sufficient range of ^{147}Sm/^{144}Nd. Whole rock studies have an advantage over mineral studies in that the scale of sampling can be much larger (meters or kilometers instead of millimeters or centimeters), so the possibility of postcrystallization isotopic reequilibration between samples is reduced (Compston and Jeffrey 1959). However, a problem with whole rock dating is in determining if the rocks are both isochronous and had the same initial ^{143}Nd/^{144}Nd ratio. Variations in initial ratios for suites of young lavas from a single volcano have been found (e.g. Chen and Frey 1983), and suggest that an assumption of a well-defined initial ratio for many suites of rocks would be difficult to defend. This problem can often be overcome for Rb-Sr dating by measuring rocks with sufficiently high ^{87}Sr/^{86}Sr that the uncertainty in the initial ratio is unimportant (e.g. Zartman 1964). However, it is particularly problematic for Sm-Nd whole rock dating because the relatively small range of ^{143}Nd/^{144}Nd in most rock suites means that any differences in the initial ratio that are larger than the analytical uncertainty could affect the age substantially. If a rock suite has variable initial ratios, and especially if the Sm/Nd ratios and the initial ratios are correlated, as is often the case, both the determined age and the initial ratio may be seriously in error (see Faure 1977, for examples of fictitious Rb-Sr "isochrons"). As an example, rocks of very nearly the same age (± 15 m.y.) from the Peninsular Ranges batholith of southern California exhibit variations in initial ^{143}Nd/^{144}Nd of some 0.14%, equal in magnitude to the total range of ^{143}Nd/^{144}Nd expected along a typical isochron (DePaolo 1981 b). Furthermore, the ^{143}Nd/^{144}Nd initial ratios are roughly correlated with ^{147}Sm/^{144}Nd in the rocks, so that the data scatter about an apparent "isochron" of 1.7 b.y., even though the crystallization age is about 0.1 b.y.! This problem may be less severe in older (Archean) rock suites where smaller ranges of initial ^{143}Nd/^{144}Nd are expected (see Chap. 8). Clearly, any scatter about a whole rock isochron must be critically assessed, since in could be an indication that both the determined age and initial ^{143}Nd/^{144}Nd are in error (cf. Chauvel et al. 1985).

2.2 Applications

There have been numerous applications of the ^{147}Sm-^{143}Nd decay to the determination of absolute ages of meteorites and ancient terrestrial and lunar rocks. Hamilton et al. (1977) have determined a whole rock "isochron" for greenstone belt volcanics from Rhodesia (Fig. 2.2). The rocks have been affected by low-temperature metamorphism and had previously proven nearly impossible to date reliably by Rb-Sr due presumably to redistribution of Rb

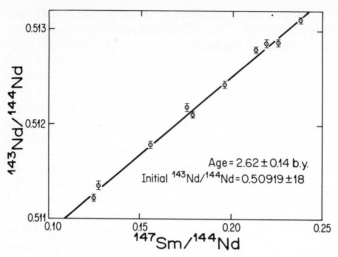

Fig. 2.2. Whole rock isochron for metamorphosed basalts from Rhodesia (Hamilton et al. 1977)

and Sr during metamorphism. The measurements by Hamilton et al. (1977) show some scatter about the best-fit line, outside of experimental uncertainty, but nevertheless define a moderately precise age. The samples are from widely separated localities, so the scatter about the line could be due to differences in initial ^{143}Nd/^{144}Nd in the rocks or to slight differences in their ages or both. These data were interpreted by the authors to show that Sm-Nd isotopes are less vulnerable to disturbance during mild metamorphism than are Rb-Sr isotopes.

An example of a mineral isochron determined on a terrestrial rock is shown in Fig. 2.3 (DePaolo and Wasserburg 1979a). The measurements were made on the minerals of a gabbro from the lower part of the banded zone of the Stillwater intrusion in southwest Montana. The data for the three minerals and the "total rock" lie precisely on a line and yield an age of 2701 ± 8 m.y. All of the points lie within one part in 10^5 of the best-fit line. The total rock point is just the sum of the three minerals since the rock contains just three minerals. A possible problem with internal isochrons is the possibility that the "phases" analyzed may be just different mixtures of only two distinct components that contain most or all of the Sm and Nd in the rock. If that is the case, even if there has been partial reequilibration of the isotopes subsequent to crystallization, the data will define a straight line, but will indicate an age that is less than the true age. For this rock, the data cannot represent a two-component mixing line because the Nd concentration in the clinopyroxene is more than five times higher than that in either of the phases at the extreme points. This data array has been interpreted as indicating the crystallization age of the intrusion. Five other whole rock samples from other parts of the

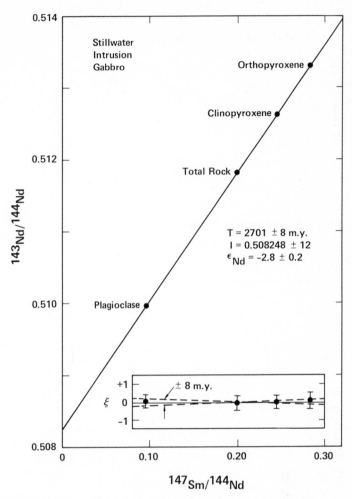

Fig. 2.3 Mineral isochron for a sample of gabbro from the Stillwater intrusion, Montana (DePaolo and Wasserburg 1979a). The *inset* shows the deviation in the y-direction of the data points from the best-fit line, in units of 10^{-4} of the ^{143}Nd/^{144}Nd ratio

intrusion also lie within analytical error of the isochron, and show that within the banded zone of the intrusion the variation of initial ^{143}Nd/^{144}Nd may have been less than ±2 parts in 10^5. Analyses of ^{87}Rb/^{86}Sr and ^{87}Sr/^{86}Sr in the same mineral separates do not define an isochron and suggest a maximum age of only about 2200 m.y. (DePaolo and Wasserburg 1979a). The results of this study demonstrate that under some conditions, redistribution of Rb and Sr can occur in a rock while Sm-Nd isotopes are undisturbed.

Figures 2.4 and 2.5 show examples of mineral isochrons determined on lunar basalts. Apollo 17 basalt 75075 (Fig. 2.4) was the first rock dated with

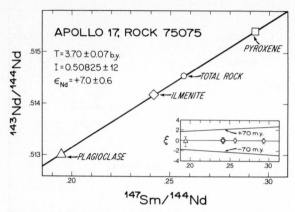

Fig. 2.4. Mineral isochron for lunar basalt 75075 (Lugmair et al. 1975a)

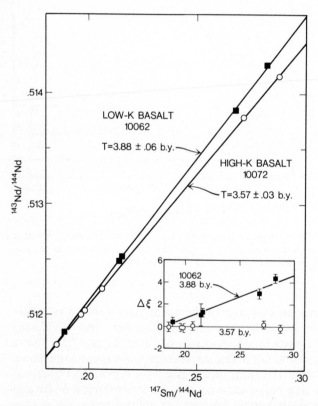

Fig. 2.5. Mineral isochrons for lunar basalts 10062 and 10072 (Papanastassiou et al. 1977). *Inset* shows the difference in the y-direction between the two isochrons in units of 10^{-4} of the $^{143}Nd/^{144}Nd$ ratio

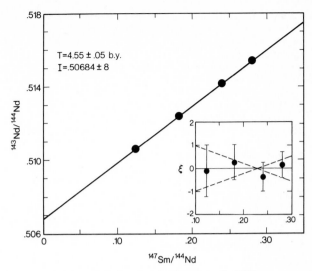

Fig. 2.6 Mineral isochron for the achondritic meteorite Moama (Hamet et al. 1978)

the Sm-Nd method (Lugmair et al. 1975a). As is generally the case, pyroxene provides a high Sm/Nd point and plagioclase a low Sm/Nd point. The total rock point in this case does not represent the sum of the minerals, since the minerals can account for only about 60% of the Sm and Nd in the total rock. Figure 2.5 shows isochrons determined on two different basalts returned from the Apollo 11 mission, superimposed on the same Sm-Nd evolution diagram (Papanastassiou et al. 1977). The age difference of 300 m.y. between these two ancient basalts can be clearly resolved. The Sm-Nd method has proven to be useful for dating low-K lunar basalts that are difficult to date with Rb-Sr because of the low Rb contents.

Figure 2.6 is a mineral isochron determined on the achondrite Moama by Hamet et al. (1978). Many achondrites are basaltic in composition and therefore are amenable to dating by Sm-Nd due to the presence of plagioclase and pyroxene which provide the necessary range in Sm/Nd. It appears that Sm-Nd systematics in meteorites may also be more resistant to disturbance than either Rb-Sr or U-Th-Pb.

It is evident from all of the examples that the typical maximum variation in $^{143}Nd/^{144}Nd$ between the minerals of a rock is about 0.5%. The determination of precise ages is therefore heavily dependent upon the capability of confidently making measurements to a precision of a few parts in 10^5.

Where comparison has been possible between Sm-Nd ages and reliable ages determined by other methods on the same rocks, the results have been in good agreement (Table 2.1). Typically, the ages agree within the stated uncertainties. It should be noted here that the decay constants of U have been measured quite precisely ($\pm 0.2\%$), but those for K, Rb, and Sm are less well

Table 2.1. Comparison of ages obtained on the same rock or formation by different methods

Sample or formation	Sm/Nd	Rb/Sr[a]	U/Pb	K/Ar	Ref.[b]
10062 (lunar basalt)	3.88 ± 0.06	3.93 ± 0.11	–	3.82 ± 0.06	1, 2
10072 (lunar basalt)	3.57 ± 0.03	3.56 ± 0.05	–	$3.62(P)^c$ $3.57(TR)^c \pm 0.04$	1, 3
Stillwater intrusion	2.701 ± 0.008	–	2.713 ± 0.004	–	4, 5
Isua metavolcanic rocks	3.75 ± 0.04		3.77 ± 0.01		6, 7

[a] Using $\lambda_{Rb} = 1.42 \times 10^{-11} \, yr^{-1}$

[b] References: 1, Papanastassiou et al. (1977); 2, Turner (1970); 3, Geiss et al. (1977); 4, DePaolo and Wasserburg (1979a); 5, Nunes (1981); 6, Hamilton et al. (1978); 7, Baadsgaard (1976); Michard-Vitrac et al. (1977)

[c] P: Plagioclase; TR: total rock

known, probably not better than $\pm 1\%$. The uncertainties in the decay constants are not included in the age uncertainties given in Table 2.1.

Chapter 3   Nd Isotopes as Tracers in Planetary Evolution

3.1 Systematics and Constructs

The principles underlying the application of the Sm-Nd radioactive system as a petrogenetic tracer are similar to those of Rb-Sr and U-Th-Pb (Faure 1977). To illustrate, an example is presented of the relationships between isotopic effects, petrogenetic processes, and time.

Consider as a starting point the formation of a planet from the solar nebula at time T_0 when the nebula has $^{143}Nd/^{144}Nd = I_0$ (Fig. 3.1).[1] The planet as a whole has a Sm/Nd ratio equal to $(Sm/Nd)_{UR}(T_0)$ which may in general be somewhat different from that of the nebula (although, as discussed in Chap. 1, it should be very close to the nebular ratio). The decay of ^{147}Sm to ^{143}Nd will cause $^{143}Nd/^{144}Nd$ in the planet $[I_{UR}(T)]$ to increase with time according to:

$$I_{UR}(T) = I_{UR}(T_0) + \left\{\frac{^{147}Sm}{^{144}Nd}\right\}_{UR} (T) \; [e^{\lambda_{Sm}(T_0-T)} - 1]$$

$$\approx I_{UR}(T_0) + 0.605 \, (Sm/Nd)_{UR} \, \lambda_{Sm}(T_0 - T) \; , \tag{3.1}$$

where T is measured backward from today (i.e., T = Age) and $(Sm/Nd)_{UR}$ is the weight ratio. The $^{143}Nd/^{144}Nd$ ratio in UR increases almost linearly with time along the trajectory $I_{UR}(T)$ at a rate (slope) proportional to the Sm/Nd in the planet. Any reservoir that retains the same constant Sm/Nd ratio throughout the history of the planet (except for ^{147}Sm decay) and whose $^{143}Nd/^{144}Nd$, therefore, evolves along a simple straight-line trajectory, is here termed a "uniform reservoir" (UR).

Let it be assumed that initially the interior of the planet is compositionally uniform and has the same Sm/Nd ratio throughout. At time T_1, a portion of the interior of the planet is partially melted. The magma thus formed coalesces, rises, and is erupted at the surface to form a segment of volcanic "crust" on the planet. Insofar as the crust is enriched in elements that were preferentially incorporated into the magma during the melting process, there will be a zone in the interior of the planet that will be depleted in these "magmaphile" elements relative to the surrounding, pristine "UR" material.

---

[1] The use of the letter I to represent $^{143}Nd/^{144}Nd$ is a reminder that "initial" $^{143}Nd/^{144}Nd$ ratios of igneous rocks must be used to trace the isotopic evolution of major reservoirs in the planet.

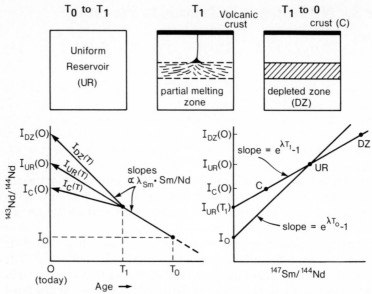

Fig. 3.1. The formation by melting of reservoirs with high and low Sm/Nd ratios from an initial homogeneous reservoir, and the change of the ^{143}Nd/^{144}Nd ratio with time in these reservoirs

If Sm or Nd is preferentially partitioned into the magma, then the depleted zone as well as the crust will have a Sm/Nd ratio that is different from UR. The evolution of ^{143}Nd/^{144}Nd subsequent to T_1 in the crust, UR, and the depleted zone (DZ) is shown in Fig. 3.1. For this example, a case has been chosen where the crust is enriched in Nd relative to Sm. This corresponds to the typical Sm-Nd partitioning in the generation of basaltic magmas in the earth (Chap. 4). Subsequent to T_1, ^{143}Nd/^{144}Nd in the crust $I_C(T)$, evolves more slowly than in UR due to the lower Sm/Nd of the crust. In the depleted zone, ^{143}Nd/^{144}Nd evolves more rapidly than in UR because of its higher Sm/Nd, as shown by the curve $I_{DZ}(T)$.

For this example the crustal segment would be said to have an age = T_1 and an initial ^{143}Nd/^{144}Nd of I_1. The difference in ^{143}Nd/^{144}Nd today between the crust and the bulk planet is:

$$\Delta I_C(0) = I_C(0) - I_{UR}(0) = \left[\left(\frac{^{147}Sm}{^{144}Nd} \right)_C - \left(\frac{^{147}Sm}{^{144}Nd} \right)_{UR} \right] [e^{\lambda T_1} - 1]$$

$$\approx 0.605 \, (Sm/Nd_C - Sm/Nd_{UR}) \lambda_{Sm} T_1 \ . \tag{3.2}$$

A similar equation can be written for ΔI_{DZ}. The isotopic differences are just proportional to the differences in Sm/Nd and the time T_1. In general, isotopic differences of a given magnitude can be caused by subtle differences in chemical composition of the reservoirs (i.e., differences of Sm/Nd) that have existed over most of the planet's history, or large chemical differences that formed recently. Also, by material balance:

$$\Delta I_{DZ}(T) M_{DZ} [Nd]_{DZ} + \Delta I_C(T) M_C [Nd]_C = 0 \ , \tag{3.3}$$

where M is the mass of the reservoir and [Nd] is the concentration of Nd. The relative magnitudes of ΔI_{DZ} and ΔI_C are therefore a function of the relative masses of the two reservoirs and their Nd concentrations. Insofar as there exists an understanding of the behavior of Sm and Nd during the petrogenetic processes responsible for the formation of the reservoirs (e.g. partial melting), the isotopic effects can be related to those processes and time in this manner.

New crustal layers can form at various times throughout the history of the planet. If the magmas are always melted from pristine UR material, the *initial* ^{143}Nd/^{144}Nd values of all segments of the crust will lie along the curve $I_{UR}(T)$. These initial values will therefore give an indication of the Sm/Nd of the bulk planet. On the other hand, if new crustal layers are sometimes derived from previously "depleted" zones, their initial ^{143}Nd/^{144}Nd will lie above the $I_{UR}(T)$ curve. In this case we would expect the I_C values of crustal rocks to lie on or above $I_{UR}(T)$, and since the isotopic differences between depleted zones and UR material grow with time, we would expect the I_C values to "fan out" at more recent times. Still another possibility is that the planetary interior is fluid and well mixed, so that any depleted zone has a finite lifetime before being totally remixed with surrounding material. In this case, if the total mass of depleted material is small in comparison to the mass of the planet, and if the mixing times are short, the isotopic evolution in the interior will nowhere depart significantly from the $I_{UR}(T)$ curve. The difference between this case and that where crust is always derived from pristine UR material would be subtle as far as the isotopic observations are concerned, but would be very important with regard to the thermal history of the planet and the degree to which the planet's interior might be stirred by convection and degassed.

If the planet's crust became thick enough, the base of the crust could become sufficiently hot that it could be partially melted. Igneous rocks derived from the crust in this manner could have I values that lie below the $I_{UR}(T)$ curve. The displacement of the I value below $I_{UR}(T)$ [or $I_{DZ}(T)$] would be proportional to the age of the crustal material that was melted and the deviation of its Sm/Nd from that of the bulk planet (Eq. 3.1).

In general, the differentiation of a planet can be studied by determining the age and initial ^{143}Nd/^{144}Nd (I values) of rocks of a range of ages. The amount of dispersion of I values about the growth curve of the bulk planet will depend upon the degree to which the planet has been differentiated, the amount of Sm/Nd fractionation that was involved, and the degree of preservation of the differentiated reservoirs. If the planet becomes differentiated early and remains so, or if it becomes more differentiated with time (increasing variability of Sm/Nd), the dispersion in I values will increase with time. If the planet is a well-mixed system, the I values will cluster about the $I_{UR}(T)$ curve at all times.

The determination of the $^{143}Nd/^{144}Nd$ evolution curve for undifferentiated material or for the bulk planet is critical to fully interpreting the isotopic data. For instance, referring again to Fig. 3.1, if $I_{UR}(T)$ is known, then if zero-age igneous rocks were found with initial $^{143}Nd/^{144}Nd$ values of $I_C(0)$ and $I_{DZ}(0)$, it could be inferred that (1) two distinct reservoirs of great age exist in the planet, (2) *both* reservoirs are differentiated, and (3) because they have $^{143}Nd/^{144}Nd$ higher and lower than UR, the reservoirs may be complementary (as shown in Fig. 3.1), and material balance considerations (Eq. 3.3) can be used to constrain the relative masses of the reservoirs. Without knowledge of the curve $I_{UR}(T)$, however, only the first of these inferences could be made. Any relationships between the reservoirs or their possible modes of formation would remain obscure. For this reason, much emphasis has been placed on the determination of a "bulk earth" evolution curve (e.g. DePaolo and Wasserburg 1976a; Jacobsen and Wasserburg 1980a, 1984).

3.2 Comparison with Other Isotopes

The differences between Sm-Nd, Rb-Sr, and U-Pb with regard to planetary differentiation are illustrated in Fig. 3.2. The Lu-Hf system (Patchett and Tatsumoto 1980a, b; Patchett et al. 1981) is almost exactly analogous to Sm-Nd, so the description of the Sm-Nd system applies to Lu-Hf also. For U-Pb, the isotope ^{206}Pb is the product of ^{238}U decay, so the slope of an evolution line on Fig. 3.2 is proportional to $^{238}U/^{204}Pb$. This ratio is known to have been very small in the solar nebula, as evidenced by the present solar and chondritic abundances, so the evolution curve shown for the nebula (labeled SN) has a small slope. As the nebula condensed, U — which has a high condensation temperature — condensed earlier than Pb, which has a low condensation temperature (Grossman and Larimer 1974). If the earth accreted before all of the Pb condensed, or if Pb was lost by volatilization during accretion, then the earth would have formed with a U/Pb ratio higher than that of the SN. This is shown on Fig. 3.2 by the increased slope of the earth's curve ($\oplus$) following condensation and accretion (CA). Further fractionation of U from Pb probably occurred as a result of core formation (CF). The tendency for Pb to occur as a sulfide could have resulted in its entering the core in large amounts while U would have been excluded from the core. The result would be a high U/Pb silicate portion of the earth (mantle and crust), and a low U/Pb core. Because there are so many opportunities for U/Pb fractionation during the formation and early evolution of the earth, a "bulk earth" evolution curve for $^{206}Pb/^{204}Pb$ is difficult to determine. Subsequent to core formation, magmatic processes in the silicate portion can cause formation of domains of somewhat higher and lower U/Pb to form, shown by the diverging arrows, but the U/Pb fractionation in magmatic processes is apparently much less than for condensation or core formation. The kinks in the curves, corresponding to changes in U/Pb, are the times that are recorded with most sensitivity. The U-Pb

Fig. 3.2. Schematic isotopic evolution diagrams for the earth illustrating the difference between the Sm-Nd, Rb-Sr, and U-Pb systems (DePaolo 1981a). *C* continents; *UM* upper mantle; *CHUR* chondritic reservoir; ⊕ bulk earth; *SN* solar nebula; *CA* condensation and accretion; *CF* core formation

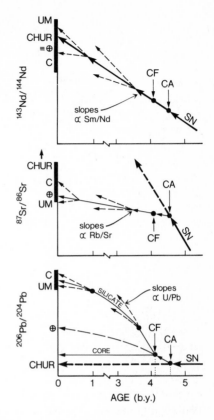

system gives very precise information on the time of formation of the earth and/or core formation (Patterson 1956), but is less sensitive to the time of separation of the crust from the mantle, because little change in U/Pb is involved in the latter process.

The situation for Rb-Sr is somewhat different (Fig. 3.2). Rb is substantially more volatile than Sr under the conditions prevailing in the solar nebula, and the earth apparently inherited a Rb/Sr ratio some ten times or more smaller than the sun's (Fig. 1.1). But, fractionation during core formation is less likely, so Rb-Sr precisely fixes the time of the earth's formation. Subsequent magmatic processes have also fractionated Rb and Sr by large factors, so that information on the age of the crust is also given by Rb-Sr. The present $^{87}Sr/^{86}Sr$ of crust and upper mantle presumably straddle the earth value, but the exact earth value is difficult to determine because of the large fractionation during the earth's formation, and because essentially all materials at the earth's surface have been affected by magmatic fractionations at some time (see Hurley 1968). The evolution curves for $^{87}Sr/^{86}Sr$ are almost straight lines as for $^{143}Nd/^{144}Nd$ because the half-life of ^{87}Rb is so long (50 b.y.).

The Sm-Nd system (Fig. 3.2) is different because both Sm and Nd are refractory in terms of the nebular condensation sequence and probably con-

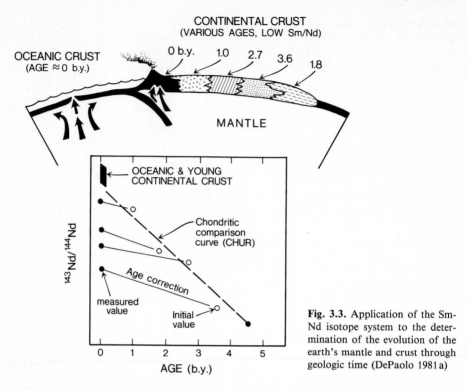

Fig. 3.3. Application of the Sm-Nd isotope system to the determination of the evolution of the earth's mantle and crust through geologic time (DePaolo 1981a)

densed quantitatively from the solar nebula early. Furthermore, neither would have entered the core, so there was no fractionation of Sm from Nd at the early stages of planetary evolution. However, substantial fractionation of Sm from Nd does occur in subsequent magmatic processes in the silicate portion of the planet. Consequently, the Sm-Nd system gives no information on the age of the earth, but provides an excellent means to study the magmatic differentiation of the planet without complications related to its original formation. A good estimate of the present-day ^{143}Nd/^{144}Nd for the total earth is provided by measurements on the chondritic meteorites (Jacobsen and Wasserburg 1980a; see Chap. 6).

The application of the Sm-Nd method to the particular problems of the history and structure of the earth's mantle is illustrated in Fig. 3.3 (DePaolo 1981a). Measurements of the ^{143}Nd/^{144}Nd ratio in oceanic volcanic rocks, all of which are relatively young in comparison to the age of the earth, and in some young continental igneous rocks, give an indication of the value and the variability of this ratio in the upper mantle today. The young basaltic rocks represent solidified magmas that recently came from the upper mantle. Older rocks are present in the continental crust, and can be used to determine the ^{143}Nd/^{144}Nd in the mantle at various times in the past. The assumption is that all of the rock materials making up the continents were at one time derived from the mantle as magma, which solidified and remained near the surface because of the low density. As shown in Fig. 3.3, the ^{143}Nd/^{144}Nd ratio is measured in a rock sample and then corrected for the rock's age back to an "initial" value, which represents the ^{143}Nd/^{144}Nd ratio the rock had at the time it came from the mantle. The slope of the age-correction vector is proportional to the Sm/Nd ratio measured in the rock, as in Fig. 3.2. The initial ratios can then be compared to the ^{143}Nd/^{144}Nd growth in a chondritic environment (called "CHUR" for Chondritic Uniform Reservoir). Deviations from this curve are indicative of chemical differentiation in the mantle, and can be interpreted in terms of models of earth evolution. Recovering information about the mantle from rocks in the continental crust can be difficult, because some igneous rocks do not represent magmas derived from the mantle, but rather appear to be melted from the crust itself. Such rocks can give no information about the mantle. On the other hand, the ^{143}Nd/^{144}Nd initial ratios can often be used to identify magmas derived from the crust and thereby used to study the structure and evolution of the unexposed lower levels of the crust (e.g. Farmer and DePaolo 1983).

3.3 Notation

3.3.1 Samarium-Neodymium

Variations of the ratio ^{143}Nd/^{144}Nd are quite small in nature so their representation in terms of deviations from a "standard" value provides both a more meaningful and a less cumbersome set of parameters (Fig. 3.4). Following the emphasis on "bulk planet" evolution curves for ^{143}Nd/^{144}Nd, the reference values of ^{143}Nd/^{144}Nd are those that would be found in a reservoir that has had a Sm/Nd ratio equal to that of the average chondritic meteorite for all time. This standard reservoir has been referred to as CHUR (Chondritic Uniform Reservoir, DePaolo and Wasserburg 1976a) and its ^{143}Nd/^{144}Nd ratio at any time T billion years into the past is given by:

$$^{143}\text{Nd}/^{144}\text{Nd}_{\text{CHUR}}(T) = {}^{143}\text{Nd}/^{144}\text{Nd}_{\text{CHUR}}(0)$$
$$- {}^{147}\text{Sm}/^{144}\text{Nd}_{\text{CHUR}}(0) \, [e^{\lambda_{\text{Sm}} T} - 1] \ , \qquad (3.4)$$

where ^{143}Nd/^{144}Nd$_{\text{CHUR}}$(0) and ^{147}Sm/^{144}Nd$_{\text{CHUR}}$(0) are the values in CHUR today.

In the normalization used in this text (see Chap. 1) the parameters are:

$$^{143}\text{Nd}/^{144}\text{Nd}_{\text{CHUR}}(0) = 0.511847$$

$$^{147}\text{Sm}/^{144}\text{Nd}_{\text{CHUR}}(0) = 0.1967$$

$$\lambda_{\text{Sm}} = 0.00654 \, \text{b.y.}^{-1}$$

$$(\equiv 6.54 \times 10^{-12} \, \text{yr}^{-1}) \ .$$

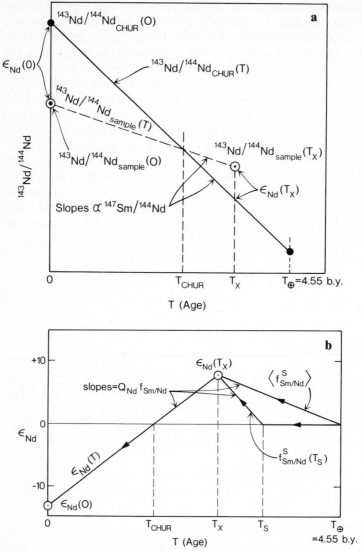

Fig. 3.4a, b. Illustration of the various neodymium isotopic evolution parameters discussed in the text

The $^{143}Nd/^{144}Nd$ of any rock sample at time T is represented by the parameter $\varepsilon_{Nd}(T)$ which is defined:

$$\varepsilon_{Nd}(T) = 10^4 \left[\frac{^{143}Nd/^{144}Nd_{Sample}(T) - {}^{143}Nd/^{144}Nd_{CHUR}(T)}{^{143}Nd/^{144}Nd_{CHUR}(T)} \right], \qquad (3.5)$$

where $\varepsilon_{Nd}(T)$ indicates the deviation of the $^{143}Nd/^{144}Nd$ value of the sample from that of CHUR in units of parts in 10^4. Since $T = 0$ is used to represent the present time, the $\varepsilon_{Nd}(0)$ of any sample refers to the $^{143}Nd/^{144}Nd$ ratio measured in the laboratory. If the sample is an igneous rock and T_x is the crystallization age, $\varepsilon_{Nd}(T_x)$ is referred to as the "initial ε_{Nd} value" of the rock, and can be likewise attributed to the magma source at time T_x.

A simple expression for the time dependence of ε_{Nd} can be derived by substituting Eq. (3.1) and the following into Eq. (3.2):

$$^{143}Nd/^{144}Nd_{Sample}(T) = {}^{143}Nd/^{144}Nd_{Sample}(0)$$
$$- {}^{147}Sm/^{144}Nd_{Sample}(0)\,[e^{\lambda_{Sm}T} - 1] \ . \qquad (3.6)$$

This gives:

$$\varepsilon_{Nd}(T) = 10^4 \left[\frac{{}^{143}Nd/^{144}Nd_{Sample}(0) - {}^{143}Nd/^{144}Nd_{CHUR}(0)}{{}^{143}Nd/^{144}Nd_{CHUR}(T)} \right]$$
$$- 10^4 \left[\frac{[{}^{147}Sm/^{144}Nd_{Sample}(0) - {}^{147}Sm/^{144}Nd_{CHUR}(0)]\,[e^{\lambda_{Sm}T} - 1]}{{}^{143}Nd/^{144}Nd_{CHUR}(T)} \right] .$$
$$(3.7)$$

This can be easily converted to the following:

$$\varepsilon_{Nd}(T) = \varepsilon_{Nd}(0) - \frac{10^4\,[{}^{147}Sm/^{144}Nd_{CHUR}(0)]\,[e^{\lambda_{Sm}T} - 1]}{{}^{143}Nd/^{144}Nd_{CHUR}(0)}\,f_{Sm/Nd}$$
$$\times \frac{{}^{143}Nd/^{144}Nd_{CHUR}(0)}{{}^{143}Nd/^{144}Nd_{CHUR}(T)} \ , \qquad (3.8)$$

where:

$$f_{Sm/Nd} = \frac{{}^{147}Sm/^{144}Nd_{Sample} - {}^{147}Sm/^{144}Nd_{CHUR}}{{}^{147}Sm/^{144}Nd_{CHUR}} \ . \qquad (3.9)$$

By making the approximation $\lambda_{Sm}T \cong e^{\lambda_{Sm}T} - 1$ and ignoring the factor in the second set of brackets, which is close to unity for any $T \le 4.5$ b.y., the expression for $\varepsilon_{Nd}(T)$ simplifies to:

$$\varepsilon_{Nd}(T) = \varepsilon_{Nd}(0) - Q_{Nd}f_{Sm/Nd}T \ , \qquad (3.10)$$

where:

$$Q_{Nd} = \frac{10^4 \cdot \lambda_{Sm} \cdot {}^{147}Sm/^{144}Nd_{CHUR}(0)}{{}^{143}Nd/^{144}Nd_{CHUR}(0)} \ , \qquad (3.11)$$

is a constant. Using the parameters given above $Q_{Nd} = 25.13$ b.y.$^{-1}$.

The expression given here (Eq. 3.10) provides a simple description of the causes of variations of the parameter ε_{Nd}. The magnitude of the ε_{Nd} value depends on the product of time and a chemical fractionation parameter ($f_{Sm/Nd}$). Large fractionation and long times produce large changes in ε_{Nd}, small fractionation relative to CHUR, and small amounts of time lead to only small shifts of ε_{Nd}. As an example of the use of Eq. (3.10), a typical crustal rock like a granodiorite might have $f_{Sm/Nd} = -0.4$. If the granodiorite formed at time T with $\varepsilon_{Nd}(T) = 0$, then, substituting into Eq. (3.10), we obtain for the present-day $\varepsilon_{Nd}(0)$ value:

$$\varepsilon_{Nd}(0) = -10.05\ T\ .$$

Hence, a 1 b.y.-old granodiorite that formed with ε_{Nd} (1 b.y.) $= 0$ would have a present-day $\varepsilon_{Nd}(0) = -10$; a 2 b.y.-old granodiorite would have $\varepsilon_{Nd}(0) = -20$, etc. Another way of expressing Eq. (3.10) gives explicitly the rate of change of $\varepsilon_{Nd}(T)$:

$$\frac{d\varepsilon_{Nd}(T)}{dT} = -Q_{Nd}f_{Sm/Nd}\ . \tag{3.12}$$

The simplified Eq. (3.10) is fairly accurate. The error is given by: $0.06\ T^2 f_{Sm/Nd}$. Hence, $\varepsilon_{Nd}(T)$ calculated from $\varepsilon_{Nd}(0)$ by Eq. (3.10) for a rock of age 4.5 b.y. with $f_{Sm/Nd} = 0.4$ would be in error by 0.5 ε units. For a more typical case of T = 2 b.y., the calculated $\varepsilon_{Nd}(T)$ would be in error by only 0.1 ε units, which is substantially smaller than the analytical uncertainty of $\pm 0.5\ \varepsilon$ units.

For an igneous rock, $\varepsilon_{Nd}(T_x)$ provides information about the magma source. A rock that has $\varepsilon_{Nd}(T_x) = 0$, for instance, could be inferred to have been derived from (i.e., the magma was derived from) a reservoir within the earth that had the chondritic value of Sm/Nd during the time period from the age of the earth $(T_\oplus)$ to T_x. A rock with $\varepsilon_{Nd}(T_x) > 0$ would be inferred to have been derived from a reservoir that had Sm/Nd greater than the chondritic ratio for at least some portion of the interval $T_\oplus$ to T_x. The Sm/Nd ratio of the magma source can be denoted $f^s_{Sm/Nd}(T)$ with the added superscript to distinguish it from the value observed in the rock. The average value of $f^s_{Sm/Nd}(T)$ as indicated by a certain value of $\varepsilon_{Nd}(T_x)$ is given by:

$$\{f^s_{Sm/Nd}\} = -\frac{1}{T_\oplus - T_x}\int_{T_\oplus}^{T_x} f^s_{Sm/Nd}(\tau)d\tau = \frac{\varepsilon_{Nd}(T_x)}{Q_{Nd}(T_\oplus - T_x)}\ . \tag{3.13}$$

A simple two-stage history for a magma source is illustrated in Fig. 3.4, where from $T_\oplus$ to T_s the source had $f^s_{Sm/Nd} = 0$, and from T_s to T_x it had $f^s_{Sm/Nd}(T_s)$. In this type of two-stage model, which can often be useful, T_s represents a "model age" of the magma source. In many cases such an age can be estimated from other geochronological relationships, especially in the particular case of large crustal provinces of known age. From this model age, the model $f^s_{Sm/Nd}(T_s)$ is calculated from:

$$f^s_{Sm/Nd}(T_s) = \frac{\varepsilon_{Nd}(T_x)}{Q_{Nd}(T_s - T_x)}\ . \tag{3.14}$$

With the estimate of $f^s_{Sm/Nd}(T_s)$ one can calculate the Sm/Nd fractionation factor $a_{Sm/Nd}(T_s)$ for Sm/Nd of the rock relative to that of the source:

$$\alpha_{Sm/Nd}(T_s) = \frac{1 + f_{Sm/Nd}}{1 + f^s_{Sm/Nd}(T_s)} \ . \tag{3.15}$$

If the average value for the source is used:

$$\alpha_{Sm/Nd}(T_\oplus) = \frac{1 + f_{Sm/Nd}}{1 + f^s_{Sm/Nd}(T_\oplus)} \ . \tag{3.16}$$

The fractionation factor $\alpha_{Sm/Nd}$ is of particular importance because it can be modelled in terms of the petrogenetic processes responsible for the formation of the rock. For a magma melted from a particular source rock, $\alpha_{Sm/Nd}$ will depend upon the mineralogy of the source rock and the degree of melting attained before the magma separated from the solid residue (Chap. 4).

The time corresponding to the point where the evolution curve of the sample $-$ $^{143}Nd/^{144}Nd_{Sample}(T)$ intersects the CHUR curve is termed T_{CHUR} (Fig. 3.4). This model age can provide a rough characterization of the age of a rock or magma source. T_{CHUR} can be calculated from the equation:

$$T_{CHUR} = \lambda_{Sm}^{-1} \ln \left[1 + \frac{^{143}Nd/^{144}Nd_{sample}(0) - {}^{143}Nd/^{144}Nd_{CHUR}(0)}{^{147}Sm/^{144}Nd_{sample}(0) - {}^{147}Sm/^{144}Nd_{CHUR}(0)} \right] \ .$$
$$\tag{3.17}$$

For the ε-notation, it is approximately (Fig. 3.4):

$$T_{CHUR} \approx \frac{\varepsilon_{Nd}(0)}{Q_{Nd}f_{Sm/Nd}(0)} \ . \tag{3.18}$$

3.3.2 Other Isotopic Systems

This notation can also be extended to any system where the increase in the isotopic ratio of the daughter element in the reference reservoir is a few percent or less over the history of the earth. This condition is satisfied for the Rb-Sr isotopic system, as well as Lu-Hf, but is not applicable for U-Pb. For Lu and Hf, both of which are cosmochemically nonvolatile, the Lu/Hf ratio for the reference reservoir can be chosen to be that of average chondrites, as for Sm/Nd (Patchett and Tatsumoto 1980a, b; Patchett et al. 1981). Thus ε_{Hf} is defined:

$$\varepsilon_{Hf}(T) = 10^4 \left[\frac{^{176}Hf/^{177}Hf_{Sample}(T)}{^{176}Hf/^{177}Hf_{CHUR}(T)} - 1 \right] \tag{3.19}$$

and

$$^{176}Hf/^{177}Hf_{CHUR}(T) = {}^{176}Hf/^{177}Hf_{CHUR}(0)$$
$$- {}^{176}Lu/^{177}Hf_{CHUR}(0) \ [e^{\lambda_{Lu}T} - 1] \ , \tag{3.20}$$

where

$$^{176}\text{Hf}/^{177}\text{Hf}_{\text{CHUR}}(0) = 0.28295$$

$$^{176}\text{Lu}/^{177}\text{Hf}_{\text{CHUR}}(0) = 0.0334$$

$$\lambda_{\text{Lu}} = 1.94 \times 10^{-11}\ \text{yr}^{-1}\ .$$

Also:

$$f_{\text{Lu/Hf}}(T) = \frac{^{176}\text{Lu}/^{177}\text{Hf}_{\text{Sample}}(T)}{^{176}\text{Lu}/^{177}\text{Hf}_{\text{CHUR}}(T)} - 1\ ; \tag{3.21}$$

$$\varepsilon_{\text{Hf}}(T) = \varepsilon_{\text{Hf}}(0) - Q_{\text{Hf}}f_{\text{Lu/Hf}}T\ , \tag{3.22}$$

where:

$$Q_{\text{Hf}} = \frac{10^4 \cdot \lambda_{\text{Lu}} \cdot {}^{176}\text{Lu}/^{177}\text{Hf}_{\text{CHUR}}(0)}{^{176}\text{Hf}/^{177}\text{Hf}_{\text{CHUR}}(0)} = 22.90\ \text{b.y.}^{-1}\ . \tag{3.23}$$

For Rb-Sr the definition of a meaningful reference or "bulk planet" evolution curve is more problematic. Because the Rb/Sr ratio of the crust-mantle system of the earth is clearly much different from the chondritic ratio ($\text{Rb/Sr}_\oplus \approx 0.1\ \text{Rb/Sr}_{\text{CHUR}}$) there is no way to choose a bulk earth Rb/Sr *a priori*. However, because in many young oceanic basalts $^{143}\text{Nd}/^{144}\text{Nd}$ and $^{87}\text{Sr}/^{86}\text{Sr}$ are closely correlated (Chap. 5), it is possible to estimate the earth's Rb/Sr ratio. It should be kept in mind, though, that the estimate is based on an interpretation of the meaning of the correlation (Chap. 5) and hence may be less reliable than the estimates for Sm/Nd and Lu/Hf. Because the reference reservoir for Rb-Sr is clearly *not chondritic* in terms of the Rb/Sr ratio, it has been referred to simply as a uniform reservoir (UR). The parameters are defined as follows:

$$\varepsilon_{\text{Sr}}(T) = 10^4 \left[\frac{^{87}\text{Sr}/^{86}\text{Sr}_{\text{Sample}}(T)}{^{87}\text{Sr}/^{86}\text{Sr}_{\text{UR}}(T)} - 1 \right] \tag{3.24}$$

$$^{87}\text{Sr}/^{86}\text{Sr}_{\text{UR}}(T) = {}^{87}\text{Sr}/^{86}\text{Sr}_{\text{UR}}(0) - {}^{87}\text{Rb}/^{86}\text{Sr}_{\text{UR}}(0)\ [e^{\lambda_{\text{Rb}}T} - 1]\ , \tag{3.25}$$

where:

$$^{87}\text{Sr}/^{86}\text{Sr}_{\text{UR}}(0) = 0.7045$$

$$^{87}\text{Rb}/^{86}\text{Sr}_{\text{UR}}(0) = 0.0827$$

$$\lambda_{\text{Rb}} = 0.0142\ \text{b.y.}^{-1}$$

$$f_{\text{Rb/Sr}}(T) = \frac{^{87}\text{Rb}/^{86}\text{Sr}_{\text{Sample}}(T)}{^{87}\text{Rb}/^{86}\text{Sr}_{\text{UR}}(T)} - 1 \tag{3.26}$$

$$\varepsilon_{\text{Sr}}(T) = \varepsilon_{\text{Sr}}(0) - Q_{\text{Sr}}f_{\text{Rb/Sr}}T\ , \tag{3.27}$$

where:

$$Q_{\text{Sr}} = 16.70\ \text{b.y.}^{-1}\ .$$

For both Lu-Hf and Rb-Sr, model ages (T_{CHUR} for Lu-Hf and T_{UR} for Rb-Sr) and magma source parameters can also be defined in a manner analogous to those defined above for Sm-Nd.

3.3.3 Alternative Notations

The ε_{Nd} parameter has also been called ε_{JUV} (Lugmair et al. 1975a) which refers to the fact that the parameters that define $\varepsilon_{Nd} = 0$ were originally chosen to be those of the achondrite Juvinas. Subsequently, those parameters have been shown to be almost exactly the same as the average chondritic meteorite (Jacobsen and Wasserburg 1980a). Another similar parameter has been proposed by O'Nions et al. (1977), called Δ_{Nd}:

$$\Delta_{Nd}(T) = 10^2 \left[\frac{{}^{147}Sm/{}^{144}Nd^*_{Source}(T)}{{}^{147}Sm/{}^{144}Nd_{CHUR}(T)} - 1 \right], \qquad (3.28)$$

where

$$ {}^{147}Sm/{}^{144}Nd^*_{Source}(T) = \frac{{}^{143}Nd/{}^{144}Nd_{Sample}(T) - {}^{143}Nd/{}^{144}Nd_{CHUR}(4.55)}{\exp\left[\lambda_{Sm}(4.55-T)\right] - 1}. $$

$$ (3.29) $$

This parameter is essentially equivalent to the previously defined time-averaged $f^S_{Sm/Nd}$:

$$ \Delta_{Nd}(T) = 10^2 \cdot \{f^S_{Sm/Nd}\}(T) = \frac{10^2 \varepsilon_{Nd}(T)}{Q_{Nd}(4.55-T)}. \qquad (3.30) $$

For rocks of very young age ($T \approx 0$), $\Delta_{Nd}(0) = 0.875\, \varepsilon_{Nd}(0)$. A similar parameter was defined for Rb-Sr. A parameter to describe the Sm/Nd fractionation between an igneous rock and its magma source was denoted δ_{Nd}:

$$ \delta_{Nd} = 10^3 \left[\frac{{}^{147}Sm/{}^{144}Nd_{Sample}(T)}{{}^{147}Sm/{}^{144}Nd^*_{Source}(T)} - 1 \right]. \qquad (3.31) $$

This parameter is similar to $\alpha_{Sm/Nd}(T_\oplus)$:

$$ \delta_{Nd} = 10^3 \left[\alpha_{Sm/Nd}(T_\oplus) - 1\right] = 10^3 \left[\frac{1 + f_{Sm/Nd}}{1 + f^s_{Sm/Nd}(T_\oplus)} - 1 \right], \qquad (3.32) $$

but is less general than $\alpha_{Sm/Nd}(T)$ because the source "age" is specified as $T_\oplus$.

A model age T_{ICE} has also been used by Lugmair et al. (1976) (ICE: Intercept with Chondritic Evolution curve). It is equivalent to T_{CHUR}.

Chapter 4 Igneous Processes and Nd Isotopic Variations

4.1 Partial Melting

The behavior of Sm, Nd, and the other rare-earth elements in magmatic processes can be conveniently illustrated by theoretical models of partial melting, fractional crystallization, and mixing. With regard to partial melting, the models that will be described here carry the assumption that equilibrium is always maintained between the magma and the residual solid material in the melting rock.

Magma forms when rock material is heated to a temperature above its solidus temperature. As the temperature is increased beyond the solidus temperature, the weight fraction of the rock that is transformed to the liquid state, denoted by F, increases. For any element i, the concentration in the liquid phase, C_i^l is given by (Schilling and Winchester 1967; Gast 1968; Shaw 1970):

$$C_i^l(F) = C_i^0 \frac{1}{F + (1 - F) \cdot D_i^0(F)} , \qquad (4.1)$$

where C_i^0 is the concentration in the solid rock prior to melting, and $D_i^0(F)$ is the bulk solid/liquid distribution coefficient, which is equal to $C_i^s(F)/C_i^l(F)$. The parameter $C_i^s(F)$ is the concentration in the solid material when the melt fraction is F. $D_i^0(F)$ is determined by the mineralogy of the solid residue:

$$D_i^0 = \sum_\varrho D_{i\varrho} \cdot X_\varrho(F) , \qquad (4.2)$$

where $D_{i\varrho}$ is the distribution coefficient for element i in phase ϱ relative to the liquid phase and X_ϱ is the weight fraction of phase ϱ in the residual solid that is in equilibrium with the liquid.

Values of the distribution coefficients ($D_{i\varrho}$) for rare earths between common minerals and silicate liquids are given in Table 4.1 and shown in Fig. 4.1. The values given were selected from a compilation by Arth (1976). The absolute values are somewhat uncertain, and vary with the composition of both the minerals and the liquids. The relative values, however, are considered to be accurate.

To model the rare-earth element abundances in a magma formed by equilibrium partial melting of a rock of a particular mineralogical composition (e.g. peridotite, eclogite, amphibolite), one also needs to know the mineralogy

Table 4.1. Rare earth element distribution coefficients

	OL[a]	OPX	CPX[b]	GAR[b]	PLAG	AMPH[c,d]	GAR[d]	CPX[d]
Ce	0.008	0.004	0.080	0.005	0.055	0.20	0.028	0.15
Nd	0.009	0.008	0.17	0.039	0.031	0.33	0.068	0.31
Sm	0.0095	0.016	0.26	0.15	0.020	0.52	0.29	0.50
Eu	0.010	0.021	0.28	0.25	0.20[e]	0.59	0.49	
Gd	0.0107	0.026	0.32	0.40	0.017	0.63	0.97	0.61
Dy	0.012	0.044	0.34	0.90	0.015	0.64	3.17	0.68
Er	0.014	0.068	0.32	1.60	0.014	0.55	6.56	0.65
Yb	0.016	0.100	0.29	3.00	0.013	0.49	11.5	0.62

[a] These distribution coefficients used also for spinel.
[b] Used for calculations of melting of peridotite.
[c] Amphibole coefficients are particularly problematic because of a strong dependence on composition.
[d] Used for melting and crystallization of "basalt" compositions.
[e] Dependent on oxygen fugacity.

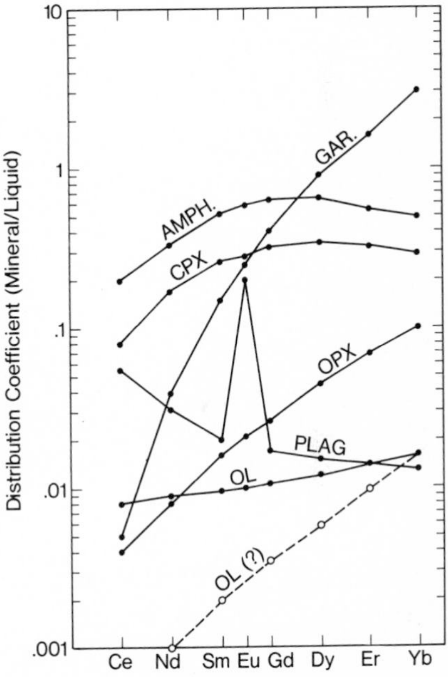

Fig. 4.1. Distribution coefficients (by weight concentration) of rare-earth elements between minerals of gabbro, eclogite, amphibolite, and peridotite and a liquid (i.e., magma) of basaltic composition. *GAR:* garnet; *AMPH:* amphibole; *CPX:* clinopyroxene; *OPX:* orthopyroxene; *PLAG:* plagioclase feldspar; *OL:* olivine. The *queried line* for OL represents a determination by McKay (1986)

of the rock and how the mineralogy, $X_\varrho(F)$, changes as the melt fraction increases. This information can be derived, with varying degrees of confidence, from available experimental phase equilibria studies (e.g. Bowen 1928; Wyllie 1979; Carmichael et al. 1974; Morse 1980). A simple example is provided by the ternary system forsterite-diopside-SiO_2 shown in Fig. 4.2a (Bowen 1914). The simple system provides a reasonable description of the melting of peri-

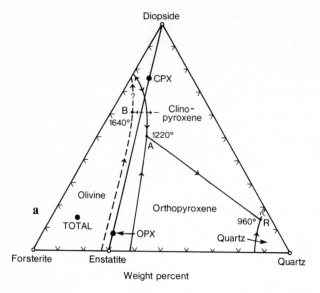

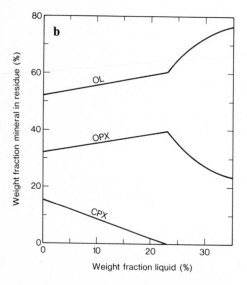

Fig. 4.2. a Melting diagram for the system forsterite(olivine)-diopside(clinopyroxene)-quartz at 1 atm. pressure. (Point B is the equivalent of A at a pressure of 20 kb.) The model rock of which **b** applies has the composition given by the *point labeled TOTAL*. The assumed pyroxene compositions are shown by the *dots*. The first liquid forms when the temperature rises to 1220 °C; it has the composition given by the *point A*. The liquid composition and the temperature remain constant as sufficient heat is added to melt 23% of the original solid rock, by which time there is no longer any clinopyroxene remaining in the solid. Upon further melting, the temperature increases and the composition of the liquid changes along the line connecting *point A* and the forsterite-quartz side of the triangular diagram. **b** The relative proportions (total solid = 100%) of *OL, OPX,* and *CPX* in the solid phase as the liquid fraction increases from 0 to 0.35

dotite like that of the earth's mantle, at low pressures in the absence of volatiles. Figure 4.2b shows the relative proportions of olivine, enstatite, and diopside in the residue as a function of melt fraction, F, for an initial rock having a composition that would plot at the point labeled "TOTAL".

For real systems there is considerably more difficulty in estimating the residual phase proportions because of the increased number of chemical components and mineral phases. Nevertheless, sufficient information exists to

Fig. 4.3. Weight fraction of mineral remaining in solid (total solid = 100%) against weight fraction of liquid (in percent) generated by partial melting for garnet peridotite, spinel peridotite, gabbro, garnet granulite, eclogite, and amphibolite. These diagrams, which were used for the construction of Figs. 4.4 through 4.9, are based on the author's assessment of a large number of experimental studies, and are simplified and approximate

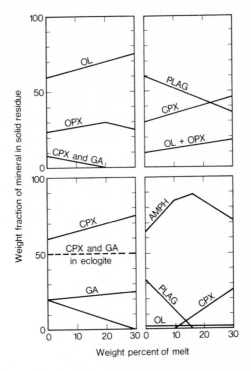

make educated guesses for most important rock types. Residual mineral proportions are given on Fig. 4.3 for rock having the composition of peridotite and basalt under different pressure conditions. Peridotite is thought to be the predominant rock type of the earth's mantle. Its mineralogy is dominated by olivine, orthopyroxene, and clinopyroxene at all upper mantle pressures, but the fourth important mineral changes from plagioclase to spinel to garnet as pressure increases (e.g. Green and Ringwood, 1967). The oceanic crust and parts of the lower continental crust have the composition of basalt. With increasing pressure, a rock of basaltic composition changes mineralogy so that it would be referred to as gabbro, garnet granulite, or eclogite. It has been suggested that eclogite may be an important rock type in the mantle also. At moderate pressure in the presence of H_2O the rock would be an amphibolite. Rare-earth element patterns *relative to the initial unmelted rock* are given in Figs. 4.4 and 4.5. Also given are the values of $\alpha_{Sm/Nd}$, which describe the change in the ratio Sm/Nd in the magma and the residual solid relative to the original rock.

   In all cases shown, $\alpha_{Sm/Nd}$ in the magma is less than unity, or, in other words, a partial melt always has Sm/Nd lower than that of the source rock. This general rule is quite important and governs much of the interpretation of Nd isotopic data. Furthermore, if the mineral garnet (and to a lesser extent clinopyroxene and amphibole) is present in a significant amount in the

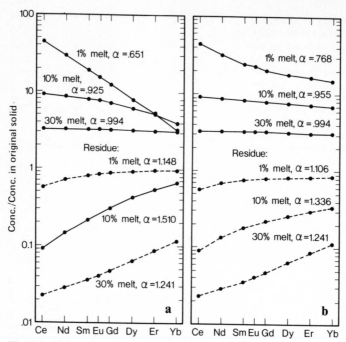

Fig. 4.4a, b. Concentration of rare-earth elements in liquid and residual solid fractions for partial melting of garnet peridotite (**a**) and spinel peridotite (**b**). The parameter a is the ratio of Sm/Nd in the liquid or residue to that in the original solid

residue, the value of a will tend to be lower than if garnet is absent. Also, $a_{Sm/Nd}$ for the liquids is most different from unity when F is small, and tends toward unity as F increases. The concentration of both Nd and Sm generally is much higher in the liquid in comparison to the original rock; the exception being when amphibole is the predominant residue mineral (Fig. 4.5).

For the residual rocks, $a_{Sm/Nd}$ is always greater than unity and *increases* as F increases. As for the liquids, $a_{Sm/Nd}$ is most different from unity if garnet is a major residual phase. Generally, Sm and Nd are depleted in the residue as the fraction of melt increases. In particular, if large fractions of melt (e.g. 30%) are removed from peridotite, Sm and Nd are almost quantitatively removed. For both liquids and residues, the maximum change in Sm/Nd that can be generated by equilibrium partial melting is about ± 50% $(0.5 \leq a_{Sm/Nd} \leq 1.5)$.

4.2 Fractional Crystallization

Another aspect of magmatic evolution that can be modeled in a simple fashion is fractional crystallization. There is considerable evidence that many

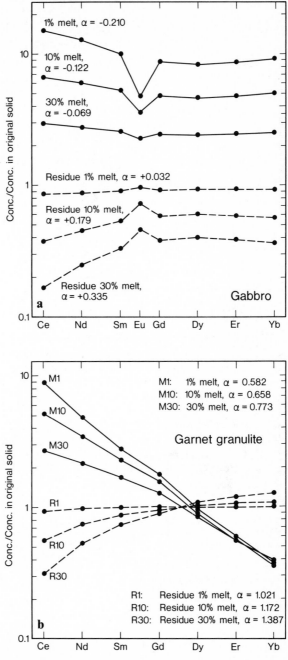

Fig. 4.5a–d. Concentration of rare-earth elements in liquid and residual solid fractions for partial melting of rocks having the chemical composition of basalt. The parameter a is the same as that of Fig. 4.4. **a** Gabbro; **b** garnet granulite; **c** eclogite; **d** amphibolite (Fig. 4c, d see page 46)

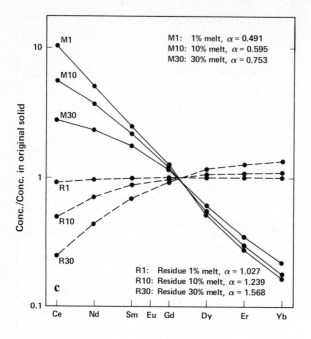

Fig. 4.5c

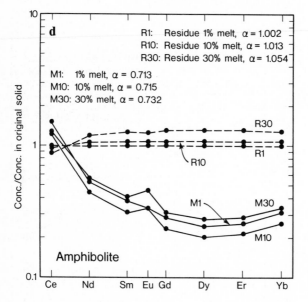

Fig. 4.5d

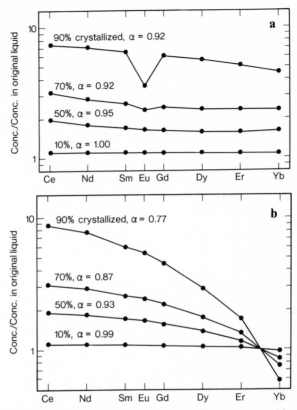

Fig. 4.6a, b. Rare-earth element enrichments caused by fractional crystallization of basaltic magma at low pressure (**a**) and high pressure (**b**). The low-pressure crystallization sequence involves initially olivine, followed by olivine and clinopyroxene, followed by clinopyroxene and plagioclase. The high-pressure crystallization involves equal amounts of garnet and clinopyroxene. The parameter a is the ratio of Sm/Nd in the liquid to that of the original liquid

magmas reach the surface only after undergoing a substantial amount of crystallization in shallow magma chambers, or even at great depth (e.g. O'Hara and Mathews 1981). For perfect fractional crystallization the concentration of an element in the liquid when a fraction F of the original liquid is left, is given by:

$$C_i^l(F) = C_i^l(0) F^{D_i^0 - 1} ,$$ (4.3)

assuming that D_i^0 is constant (Gast 1968). Figure 4.6a shows the rare-earth enrichments *relative to the initial magma* for fractional crystallization of tholeiitic basalt at low pressures ($P \gtrsim 10$ kilobars). The order of crystallization assumed was olivine, followed by pyroxene and plagioclase, but the patterns shown are not sensitive to the crystallization sequence. In general, low-pressure fractional crystallization serves mainly to enrich the residual magma

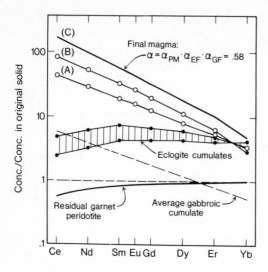

Fig. 4.7. Rare earth element concentrations (relative to original solid) in liquids generated by partial melting of garnet peridotite and then affected by fractional crystallization. *A* liquid formed by 1% partial melt of garnet peridotite (*PM*). *B* A + 50% high pressure (eclogite) fractional crystallization (*EF*). *C* A + B + 50% low pressure (gabbroic) fractional crystallization (*GF*). The patterns labelled "eclogite cumulates" and "average gabbroic cumulate" are those of the accumulated crystals formed during the stages of fractional crystallization

in all of the rare earths with slightly more enrichment of the lighter elements. Crystallization of plagioclase causes formation of a negative "europium anomaly" because Eu is concentrated in plagioclase to a greater degree than the other rare earths due to its Ca-like chemistry when reduced to the divalent state.

Figure 4.6b shows the effect of crystal fractionation at high pressures (P > 25 kilobars). The high retentivity of heavy rare earths in garnet causes depletion of Yb in contrast to enrichment for the lighter elements. The pattern generated is quite different from the low-pressure curves. However, in both cases the Sm/Nd ratio decreases only moderately even after 50–70% crystallization.

Figure 4.7 combines the partial melting and fractional crystallization effects by showing the rare-earth patterns for a magma formed by 1% partial melting of garnet peridotite followed by crystallization of half of the magma at high pressure and half of the remaining magma at low pressure. The final magma is greatly enriched in Ce (170×) relative to the magma source (dotted line) because of the further enrichment produced by extensive crystallization. Rare-earth patterns for the crystal accumulates (concentration $- D_0^i C_l^i$) formed at high and low pressure are also shown. Figure 4.7 shows clearly that the net result of magmatic processes is to produce a number of lithic reservoirs with various Sm/Nd ratios from a single original reservoir. Each reservoir (lava, cumulates, and residual peridotite) then evolves a distinct $^{143}Nd/^{144}Nd$ ratio as time progresses.

4.3 Mixing

The third important process in igneous petrogenesis is mixing. Included are mixing of different magmas and the assimilation by magma bodies of the solid surrounding them. Mixing is a special process in that the isotopic ratios in the endmembers of a mixture may be different, so that in the process of mixing one component into another, shifts in isotopic ratios as well as shifts in chemical composition occur. For example, mantle-derived magmas with one value of $^{143}Nd/^{144}Nd$ can assimilate old continental crustal rock with a different value of $^{143}Nd/^{144}Nd$ to produce igneous rocks with intermediate isotopic compositions. The properties of mixtures are also important for studies of the provenance of sedimentary rocks, where the detritus represents a mixture of materials derived from different source terrains on the continents, as well as for studies of metamorphic rocks, ocean water, hydrothermal fluids, groundwater, and rivers.

The algebra describing the isotopic composition of simple mixtures has been discussed in a number of papers (Vollmer 1976; Langmuir et al. 1978; DePaolo and Wasserburg 1979c). Using Nd as an example, the ε_{Nd} of a mixture of components α and β containing weight fractions X^α and X^β (where $X^\alpha + X^\beta = 1$), is given by:

$$\varepsilon_{Nd}^m = \frac{X^\alpha [^{144}Nd]^\alpha \varepsilon_{Nd}^\alpha + (1 - X^\alpha)[^{144}Nd]^\beta \varepsilon_{Nd}^\beta}{X^\alpha [^{144}Nd]^\alpha + (1 - X^\alpha)[^{144}Nd]^\beta} . \qquad (4.4)$$

The superscript "m" denotes the mixtures, and $[^{144}Nd]$ is the molar concentration of ^{144}Nd, which could be replaced by the weight concentration of total Nd with no significant loss of accuracy. A similar equation could be written for Sr (or Pb or Hf), which in combination with the above equation, would give the $(\varepsilon_{Nd}, \varepsilon_{Sr})$ coordinates of a mixture as a function of X^α or X^β. An alternative representation in terms of the ε_{Nd} and ε_{Sr} values is given by Vollmer (1976):

$$A\varepsilon_{Nd}^m + B\varepsilon_{Nd}^m\varepsilon_{Sr}^m + C\varepsilon_{Sr}^m + D = 0 , \qquad (4.5)$$

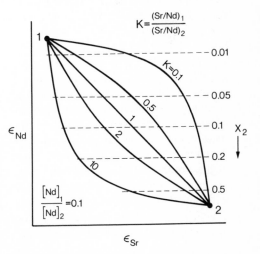

Fig. 4.8. Curves showing the isotopic composition of Nd and Sr in mixtures of the substances 1 and 2. For this calculation it is assumed that the Nd concentration in substance 2 is ten times greater than that of substance 1. The parameter X_2 is the weight fraction of substance 2 in the mixture. K is the ratio of the Sr/Nd ratios in substances 1 and 2, as shown

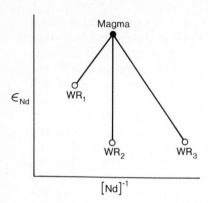

Fig. 4.9. Relationship between ε_{Nd} and the inverse of the Nd concentration for two-component mixtures involving a magma and three different wall rock (WR) endmembers

which is an equation for a hyperbola. The curve is a straight line when $B = 0$, which is the case when the following is satisfied:

$$K_{Sr/Nd}^{\alpha/\beta} = \frac{(Sr/Nd)^{\alpha}}{(Sr/Nd)^{\beta}} = 1 \ . \tag{4.6}$$

When $K_{Sr/Nd}$ is much different from unity, the mixing line is strongly curved. Curves describing the isotopic compositions for mixtures of two components are shown in Fig. 4.8.

Mixing lines between reservoirs of rock in the crust and mantle, or between the crust and the mantle, will in general be only slightly curved because Sr/Nd exhibits only a restricted variability in most rocks. Consequently, differences in mixing trajectories on an $\varepsilon_{Nd} - \varepsilon_{Sr}$ graph are most likely to reflect differences in the isotopic composition of one of the endmembers. The most notable exception to this rule is mixing between rock and ocean water, where ocean water is depleted by a factor of 10^6 in Nd relative to Sr.

The simple mixing model not only predicts the shifts in isotopic composition as the endmember proportions change, but also predicts the concurrent changes in chemical composition. For example, the denominator on the right side of Eq. (4.4) gives the Nd concentration of the mixture. Simple mixing produces well-defined relationships between elemental concentrations and isotopic compositions (Fig. 4.9).

The simple mixing models are directly applicable to any process that involves mechanical mixing. However, where chemical fractionation accompanies the mixing of isotopically different materials, the model is inadequate. An example of such a process is the assimilation of crustal rock by magma derived from the mantle. In this case the isotopic composition of the crustal rock can be radically different from that of the mantle magma, so assimilation will cause a change in the isotopic composition of the magma. At the same time, because the temperature of the crustal environment is lower than that of the magma, the magma will be cooling and crystallizing. The residual magma will be changing in chemical composition not only because of the admixture of crustal rock, but also because of fractional crystallization. Thus, for example, the fractionating

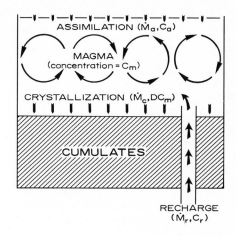

Fig. 4.10. Model of a magma chamber involving assimilation, crystallization and replenishment of the magma body (DePaolo 1985). The rates of mass transfer are denoted by $\dot{M}$ and the elemental concentrations in the transferred material are C_a, C_r, and DC_m, where D is the bulk distribution coefficient of the element between the crystallizing minerals and the magma

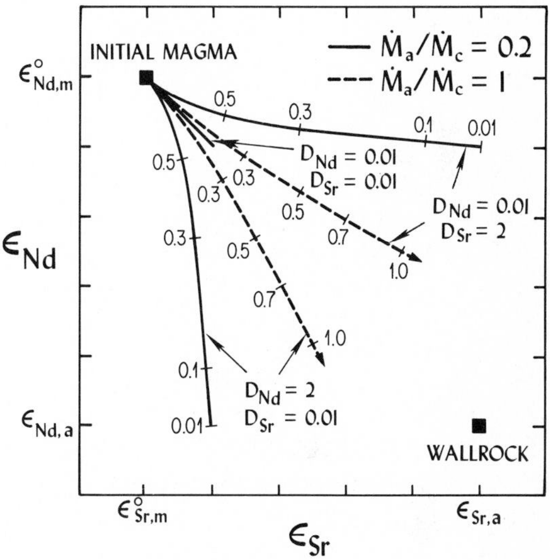

Fig. 4.11. Isotopic shifts in a magma assimilating wallrock and crystallizing, showing the different "mixing" trajectories for different distribution coefficients of Nd and Sr between the crystallizing minerals and the magma, and different ratios of the assimilation and crystallization rates (DePaolo 1981 d)

phases may have differing affinities for Sr and Nd, hence the Sr/Nd of the magma can change, which in turn modifies the mixing trajectory on the $\varepsilon_{Nd} - \varepsilon_{Sr}$ plot. More importantly, the relationship between isotopic composition and elemental concentration can be grossly different from the simple mixing case.

The mathematics associated with the assimilation-fractional crystallization (AFC) process (Fig. 4.10), and the additional possibility of adding new magma to the magma chamber as well, have been presented elsewhere (DePaolo 1981 d, 1985). Examples of mixing models are shown in Fig. 4.11 and 4.12. In the following chapters these mixing models, and the partial melting models, will be called upon as needed in the discussions of the petrogenesis of different rock suites, especially in Part III.

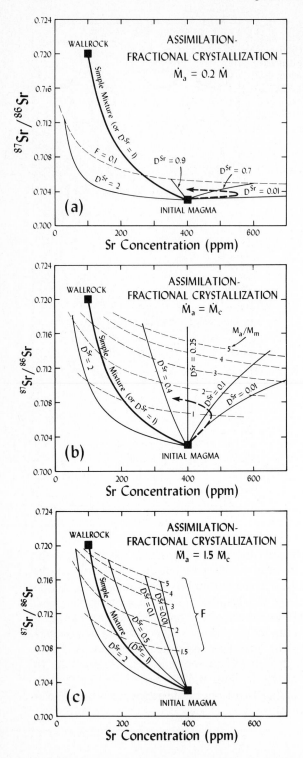

Fig. 4.12a–c. Shifts in the $^{87}Sr/^{86}Sr$ ratio and the Sr concentration in a magma assimilating wallrock and crystallizing. The different "mixing" lines reflect different bulk distribution coefficients for Sr between crystallizing minerals and the magma, and different ratios of the assimilation and crystallization rates (DePaolo 1981d). F is the ratio of the mass of the contaminated magma to that of the original magma. M_a/M_m is the ratio of the mass of assimilated wallrock to the mass of the original magma. The *heavy dashed lines* represent plausible evolution curves for a magma moving through continental crust

Part II   Nd Isotopic Variations –
A Planetary Perspective

Chapter 5 Overview of Nd Isotopic Variations

5.1 Sm-Nd Isotopic Properties of Meteorites

The chondritic meteorites probably provide the best samples of the building blocks from which the terrestrial planets were constructed (e.g. Ringwood 1975). They are primitive objects composed of condensates from the solar nebula that have suffered little further modification. As discussed in Chaps. 1 and 3, the inference that Sm and Nd condensed from the solar nebula at a high temperature implies that their relative proportions in chondritic meteorites, and indeed in all of the planets, should be the same as in the solar nebula (i.e., the sun). Because the chondrites have not been affected by magmatic processes, they provide an estimate of the bulk-earth values of Sm/Nd and $^{143}Nd/^{144}Nd$.

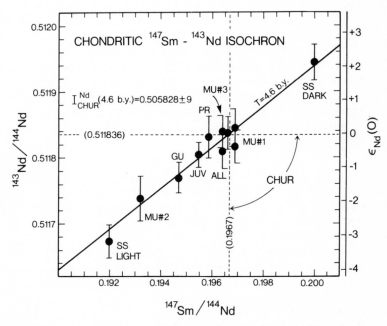

Fig. 5.1. Sm-Nd isotopic measurements of chondritic meteorites (Jacobsen and Wasserburg 1980a)

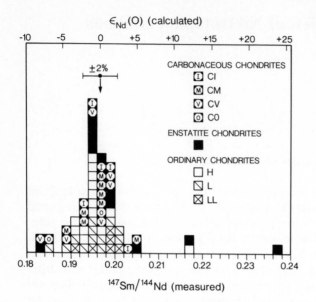

Fig. 5.2. Measurements of the $^{147}Sm/^{144}Nd$ ratio in chondritic meteorites by the isotope dilution method. The typical analytical uncertainly is $\pm 2\%$ (Jacobsen and Wasserburg 1980a)

Measurements of $^{147}Sm/^{144}Nd$ and $^{143}Nd/^{144}Nd$ were made on samples of nine chondritic meteorites by Jacobsen and Wasserburg (1980a, 1984) (Fig. 5.1). Within analytical uncertainty, the data points fit a line that corresponds to an age of about 4.6 b.y., identical to the age of the meteorites as determined by other methods (e.g. Minster et al. 1982). This indicates that the $^{143}Nd/^{144}Nd$ ratio of the solar nebula was uniform to at least about ± 0.5 ε units. Also, the range of $^{147}Sm/^{144}Nd$ values measured is small (2% excluding the two separates of St. Severin − "Light and Dark"). Although the sampling is limited, this suggests that the variability of Sm/Nd in chondrites is very small. Jacobsen and Wasserburg suggest that the variations measured are due to small sample sizes (typically 0.5 cm^3 or less), and that the chondritic material might be more homogeneous if sampled on a larger scale.

Assuming that all chondrites would lie on the isochron shown in Fig. 5.1, the present-day range of $^{143}Nd/^{144}Nd$ can be inferred from the measured range of $^{147}Sm/^{144}Nd$ ($= 0.6049 \times$ measured weight ratio Sm/Nd). This allows the more abundant data on Sm and Nd concentrations in chondrites to be used. Figure 5.2, from Jacobsen and Wasserburg (1980a), is a histogram of $^{147}Sm/^{144}Nd$ ratios measured in 64 chondrite samples by the isotope dilution method. These data are accurate to about $\pm 2\%$. Although a few samples are anomalous, the $^{147}Sm/^{144}Nd$ ratios of the vast majority lie within a few percent of the mean. If it is assumed that the 64 samples are a random, normally distributed sampling of the chondritic ratio, the calculated mean and 95% confidence limit are 0.1967 ± 0.0019.

Based on this and the data from Fig. 5.1, Jacobsen and Wasserburg (1980a) suggested that the values $^{147}Sm/^{144}Nd = 0.1967$ and $^{143}Nd/^{144}Nd = 0.511847$

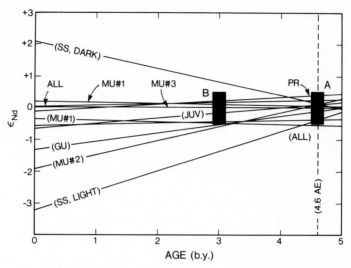

Fig. 5.3. Evolution of ε_{Nd} through time for the meteorite samples of Fig. 5.1 (Jacobsen and Wasserburg 1980a)

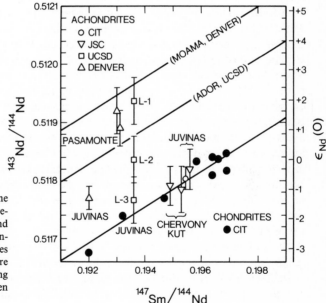

Fig. 5.4. Summary of some Sm-Nd isotopic measurements on chondritic and achondritic meteorites. Interlaboratory discrepancies are responsible for the failure of the data points to lie along a single isochron (Jacobsen and Wasserburg 1984)

be used to represent the average chondritic values (CHUR). The calculated values of $\varepsilon_{Nd}(O)$ for the various chondrite samples are also shown on Fig. 5.1 and 5.2. The evolution of $\varepsilon_{Nd}(T)$ for the samples measured by Jacobsen and Wasserburg (1980) is shown in Fig. 5.3.

The available data suggest that the chondritic, or solar system, values of $^{147}Sm/^{144}Nd$ and $^{143}Nd/^{144}Nd$ are well defined, to a level of $\pm 1\%$ and $\pm 1\varepsilon$ unit, respectively. Deviations greater than these measured in planetary materials are significant in terms of planetary evolution processes. However, interlaboratory discrepancies may be at least this large. This is illustrated by Fig. 5.4. All of the points shown, and the Moama and ADOR isochrons, should be coincident with the CIT data if there were no interlaboratory discrepancies. The measurements from different labs in this case are as much as 3 or 4 ε units different. Care must be exercised when comparing data from different laboratories, although the wider dissemination and use of standards (Wasserburg et al. 1981) has improved interlaboratory consistency.

5.2 Nd Isotopes in Mesozoic and Younger Basalts

5.2.1 Significance of Young Basalts

For a discussion of the broader characteristics of Nd isotopic variations in terrestrial rocks, it is useful to separate the samples on the basis of age. This reflects a fundamental distinction in the type of information obtained. The important separation, especially for igneous rocks, is between those younger and older than about 200 million years, the maximum age of the ocean basins. In actuality, the dividing line could be placed somewhat further into the past.

The unique characteristic of the "young" rocks is that their isotopic compositions can be interpreted in the context of their geographic and tectonic position. For example, young basalts can be subdivided into categories based on the nature of their occurrence, such as mid-ocean ridges, intraplate oceanic islands, magmatic arcs, continental margins, and continental interiors. In many cases the magmatism can be related directly to tectonic features, such as spreading ridges and subducting plates, and isotopic variations can be interpreted against a background of seismic structure, heat flow, gravity, electrical conductivity, and other characteristics of the underlying regions of the mantle or crust from which the magmas must have come. Young rocks consequently offer a rich natural laboratory for understanding and interpreting isotopic variations and igneous petrogenesis.

In contrast, very old rocks are found only in continents. If their age is greater than that of the ocean floor it becomes difficult to specify with certainty the tectonic setting in which the magmas formed, or even if the setting has an analog at present. With old rocks the chemical composition is often the only clue to their origin, especially if metamorphism has completely changed the mineralogy and texture.

Lavas having the chemical composition of basalt are particularly significant because petrologic studies indicate that the magmas they represent could form by partial melting of mantle rocks (e.g. Green and Ringwood 1967). Consequently, the isotopic compositions of basalts may directly reflect the mantle regions where the magmas formed. The basalts provide isotopic samplings of the earth's mantle and can be used to assess its structure, composition, and history.

5.2.2 Summary of Data

Figure 5.5 shows the values of initial ε_{Nd} measured in young terrestrial basalts, grouped on the basis of the tectonic setting where the basalts were erupted. Shown for comparison is the distribution of ε_{Nd} values measured on chondritic meteorites, and the distribution calculated from the measured $^{147}Sm/^{144}Nd$ (Fig. 5.2). Overall, the basalts show a large range of ε_{Nd} values, clearly indicating the existence of mantle domains with different values of Sm/Nd that have remained isolated from each other for long time periods. For example, it is known from studies of rare-earth element abundances that variations in $f_{Sm/Nd}$ are likely to be of the order $\Delta f_{Sm/Nd} = 0.1$ to 0.4 (Chap. 4). A difference of $\Delta \varepsilon_{Nd} = 10$ thus requires a time period ΔT given by (Eq. 3.7):

$$\Delta T = \frac{\Delta \varepsilon}{Q_{Nd}\Delta f_{Sm/Nd}} = 1 \text{ to } 4 \text{ b.y.}$$

In order to generate the ε_{Nd} variations observed, time periods of at least 1 b.y. are needed, but it is equally likely that these isotopic variations indicate

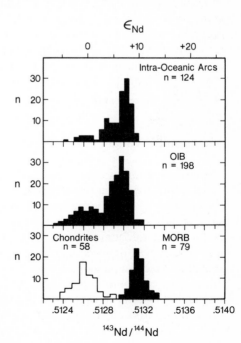

Fig. 5.5. Histogram of measured $^{143}Nd/^{144}Nd$ ratios in oceanic basalts (adapted from Morris and Hart 1983)

that separate mantle reservoirs have persisted since the time of formation of the earth 4.55 b.y. ago.

The ε_{Nd} values of oceanic basalts are the most reliable indicators of the mantle values because there is little chance that their ε_{Nd} values could be modified on the way to the surface. In contrast, the presence of thick crust with negative ε_{Nd} values in continental areas presents the distinct possibility that the basaltic magmas could become "contaminated" with the crustal Nd, so that the erupted basalt would not be representative of its source in the mantle. For the most part, oceanic basalts have ε_{Nd} values that are either the same as the chondritic meteorites or displaced toward more positive values. This implies that the Sm/Nd ratio in mantle reservoirs is either equal to or greater than the chondritic value. As shown in Chap. 4, this is just what would be expected for residual material left behind after partial melts had been extracted from the mantle. Consequently, we can interpret the data as indicating that the original mantle Sm/Nd ratio was equal to the chondritic value, and that many parts of the mantle have had magma extracted from them at some time in the past.

If the rocks are grouped by tectonic settings, it is found that each group has a somewhat narrower range of ε_{Nd} than is characteristic worldwide. Mid-ocean ridge basalts (MORB) have ε_{Nd} of about $+10$ with a standard deviation of about 1.5. These basalts show the greatest average displacement from the chondritic values and the least dispersion. Intraplate oceanic islands tend to have a large range of ε_{Nd} with some samples reaching into the range of the chondritic meteorites. Overall, the islands have ε_{Nd} that is displaced toward more negative values relative to MORB. The third major subdivision of oceanic lavas, oceanic magmatic arcs associated with subduction zones, have, for the most part, a narrow range of ε_{Nd} with an average value intermediate between the range of MORB and the average oceanic island, but also have a small segment of the population with values in the range of the chondritic meteorites.

5.2.3 Isotopic Variations as a Function of Sampling Scale

The interpretation of the ε_{Nd} data shown in Fig. 5.5 must depend on a comparison with the variations within a single magmatic province, within lavas erupted from a single volcano, and within a single pluton or lava flow. We have already seen (Fig. 2.3) that the Stillwater layered instrusion had a very uniform ε_{Nd} (± 0.2 units), so it may be expected that a single body of magma can have a well-defined ε_{Nd}, at a level comparable to or less than the analytical precision. Figure 5.6 shows ε_{Nd} data by petrologic province. Intraprovince variability ranges from no measurable variation to variations of a magnitude similar to that of the more general groupings shown in Fig. 5.5. The samples from the island of Hawaii exhibit a total spread of about 3 ε_{Nd} units (O'Nions et al. 1977; Chen and Frey 1983). The lavas from Oahu show

Fig. 5.6. Histogram of ε_{Nd} values of young basaltic lavas from five provinces. New Britain, Jan Mayen, and Oahu are oceanic. The Columbia River province and the Scottish Hebrides are continental. Data from DePaolo and Johnson (1979), Carlson et al. (1981), Carter et al. (1979), Stille et al. (1983), DePaolo and Wasserburg (1976a, b), DePaolo and Maaloe (unpublished), and O'Nions et al. (1977)

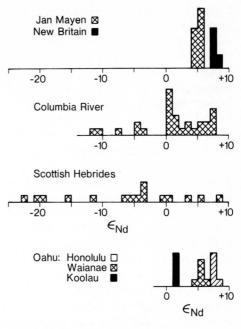

a larger range of about 7 units. In both cases, the nephelinites have the highest ε_{Nd} values, while alkali basalts and tholeiites have progressively lower values. The data from the Hawaiian lavas show that a single volcanic conduit system can access a variety of mantle reservoirs. In contrast, a suite of alkalic basalts from Jan Mayen exhibit no detectable variation of ε_{Nd}. Similarly, lavas from New Britain, ranging from basalt (48% SiO_2) to rhyolite (76% SiO_2), have ε_{Nd} with a total range of 1.4 units, only slightly greater than the analytical uncertainty. The New Britain data give an indication of how homogeneous a mantle magma source can be over a region several hundred kilometers across. In fact, the coherence of ε_{Nd} values in primitive oceanic magmatic arcs worldwide (Fig. 5.5) is noteworthy.

For the continental suites there is a greater variability of ε_{Nd}. The Columbia River province shows a range of 20 units (Carlson et al. 1981), and Tertiary basalts from the Scottish Hebrides have a range of about 28 units (Carter et al. 1978). The highest values found are in the range of the oceanic basalts, but the ε_{Nd} values extend to large negative values. This difference is almost certainly the result of some sort of contamination of the magmas by old continental rocks that have low ε_{Nd} values (Sect. 5.3), although a difference in the nature of the mantle under continents is another possibility. Most of the continental basalts have ε_{Nd} values similar to those of their oceanic counterparts. In particular, alkalic basalts erupted on continents tend to have exactly the ε_{Nd} values of similar basalts erupted in oceanic regions. For example, many alkali basalts from the western US and Africa (Allègre et al. 1981) have ε_{Nd} values of $+4$ to $+8$, similar to basalts from Hawaii and Jan Mayen.

Mantle reservoirs of this general isotopic composition are apparently common under both oceanic and continental regions. However, if the comparison is made for basalts of tholeiitic affinity, a difference is found. Almost all mid-ocean ridge basalts have $\varepsilon_{Nd} > +7$, but almost all continental and intraplate oceanic tholeiites have $\varepsilon_{Nd} < +7$. There is only minor overlap. There is a strong suggestion that the sources of the tholeiitic-type basalts are different in intraplate regions and mid-ocean ridges (see Chap. 8).

5.2.4 Implications of the Observed Isotopic Variability

A surprising aspect of the ε_{Nd} data is the rather small total range. All of the oceanic basalts are within about 12 ε units of zero (Fig. 5.5). By comparison, the range of ε_{Nd} values that might be expected, based on the partial melting models discussed in Chap. 4, is illustrated in Fig. 5.7. Petrologic studies have been interpreted as indicating that most basalts are formed by 1 to 30% melting of the mantle (e.g. Green and Ringwood 1967). Figure 5.7 shows what the ε_{Nd} evolution for residual garnet peridotite would be after a 1, 10, or 30% partial melt was removed 3.8 or 1.0 b.y. ago (assuming that the premelting peridotite had $\varepsilon_{Nd} = 0$ and chondritic Sm/Nd). For a melting event that occurred 3.8 b.y. ago, the residual peridotite would evolve very large ε_{Nd} values by the present time, especially if the melt fraction removed was in the range of a few percent to about 15%. Only for $<1\%$ melting would the present-day ε_{Nd} remain within the range of values observed in oceanic basalts. For the melting event 1.0 b.y. ago, the residue would be within the observed ε_{Nd} range for essentially any value of the melt fraction.

Although the choice of distribution coefficients, melting model, and mantle mineralogy will affect the conclusions to some degree, there are a number of possible explanations for the observed small range of ε_{Nd} values. An ob-

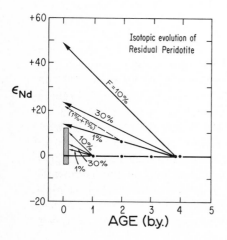

Fig. 5.7. Calculated shifts of ε_{Nd} that would evolve in the solid residue following removal of 1, 10, or 30% partial melt from a garnet peridotite at 1 and 3.8 b.y. ago

vious possibility is that oceanic basalts provide a biased view of the ε_{Nd} of the mantle. For instance, the most fractionated residues (steepest ε_{Nd} evolution curves) would be those reservoirs that have been most depleted of easily fusible materials and also heat-producing radioactive elements (K, U, Th). It is possible that such reservoirs seldom remelt to produce basalts, so that the basalts always come from the least depleted, and least fractionated, mantle regions, i.e., those with ε_{Nd} closest to zero.

If the basalts are assumed to give an unbiased representation of ε_{Nd} values in the mantle, then one or more of the following possible conclusions must be drawn:

1. Melts removed from the mantle are always less than 1% by weight of the source region (F = 1%). In this way, reservoirs with very high Sm/Nd are unlikely to be formed.
2. The regions of the mantle from which oceanic basalt magmas are melted are residual, but the ancient partial melting events that fractionated Sm/Nd occurred within the last 1 billion years.
3. Highly fractionated residual peridotite does form in the mantle as a result of removal of >1% melt, but such fractionated reservoirs are destroyed relatively quickly via mixing with less fractionated parts of the mantle and/or recycled crustal material in the mantle.

There is considerable evidence that highly fractionated reservoirs exist in the mantle. For example, some large peridotite bodies have high Sm/Nd ratios (cf. Fig. 1.1) and some basaltic rocks, which have been melted from the mantle, also have very high Sm/Nd (e.g. Zindler et al. 1979). Consequently, alternative (1) can probably be ruled out. Alternative (2) may be important, but it is almost certain that mixing (alternative 3) is largely responsible for the limited variability of ε_{Nd} observed. It is particularly noteworthy that the scatter of ε_{Nd} values in MORB is only about ± 2 units. For the 10% melting example shown in Fig. 5.7, the rate of ε_{Nd} growth in a residue relative to the original mantle before melting, is:

$$\dot{\varepsilon}_{Nd} = Q_{Nd} f_{Sm/Nd} = 13/\text{billion years} .$$

This means that it should be possible to develop reservoirs within the mantle that could produce the entire range of MORB ε_{Nd} values within about 0.3 b.y. after a partial melting event. Consequently, the coherence of the MORB ε_{Nd} data *worldwide* suggests that the lifetime of a highly fractionated reservoir in the mantle is only a few hundred million years. This implies that the upper mantle is mixed and homogenized on a relatively short time scale.

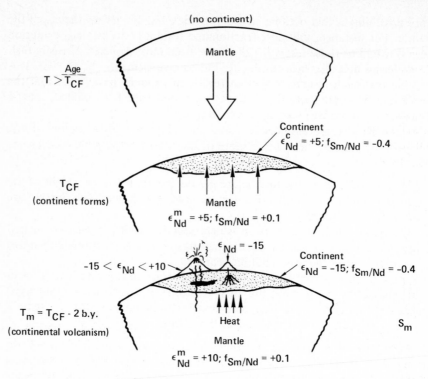

a

Note: $\epsilon_{Nd}(T_m) = \epsilon_{Nd}(T_{CF}) + Q_{Nd}f_{Sm/Nd}(T_{CF} - T_m)$

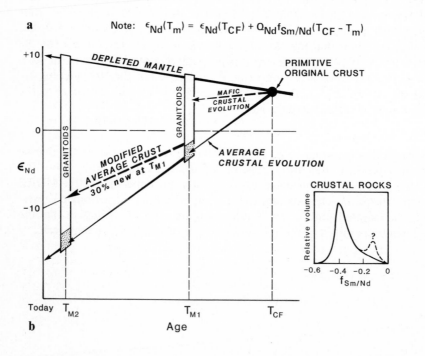

b

5.3 Nd Isotopes in Paleozoic and Precambrian Igneous Rocks

5.3.1 A Model for the Interpretation of Initial Isotopic Ratios in Ancient Terranes

Studies of Pre-Mesozoic (age ≥ 250 million years) rocks must contend with the restriction of sampling to the continents. This requires the use of a different paradigm for interpreting the data (Fig. 5.8). The fundamental assumption is that continents (i.e., continental-type crust) are formed by chemical fractionation processes from the mantle, so that at the time a segment of continental crust first forms, it has an ε_{Nd} value identical to that of the mantle. Subsequently, the ε_{Nd} values of the continent and the mantle diverge sharply because of differing values of $f_{Sm/Nd}$. For the example shown, the continent and the mantle have $\varepsilon_{Nd} = +5$ at the time the continent originally forms (T_{cf}). After 2 b.y. have passed, the ε_{Nd} of the mantle has evolved to $+10$, while that of the crust has shifted to -15. At times T_{M1} and T_{M2}, magma from the mantle is erupted through or emplaced into the continent. This magma enters the crust with an ε_{Nd} value equal to that of the mantle, but because it can assimilate rock material from the continent that has low ε_{Nd}, by the time the magma is erupted or solidified in the crust, its ε_{Nd} may be anywhere between the mantle and the crustal values, depending on the extent of its interaction with the crust. In the extreme case, the magma may be derived wholly by melting of the crust with no mantle involvement at all.

As a rule, each continent is composed of a finite number of segments, each of which was formed at a particular time during earth history (e.g. Hurley and Rand 1969; Muehlburger 1980). For example, some 70% or more of North America is composed of only three segments, or provinces, with ages of about 2.8, 1.8, and 1.3 b.y. The model illustrated in Fig. 5.8 implies that if the initial value of ε_{Nd} for the oldest rocks of any province are determined, the ε_{Nd} of the mantle at the time that particular crustal segment formed ($\equiv T_{cf}$; the crust-formation age) will be recovered. However, if the initial ε_{Nd} values are determined for rocks within the province but of age T_m, younger than T_{cf}, the mantle ε_{Nd} at T_m will not, in general, be recovered. Instead, some value between the mantle ε_{Nd} and the characteristic ε_{Nd} of the continental province at that time, will be determined. As discussed below, in some cases substantial pieces of continental crust have formed that do not have the mantle value at

◄

Fig. 5.8. a Model for the interpretation of Nd isotopic compositions of continental igneous rocks. **b** Evolution of the ε_{Nd} value in the crust and depleted mantle for continental crust that forms from the mantle at time T_{CF}. Later magmatic events occur at times T_{M1} and T_{M2}, and can result in granitic rocks that have ε_{Nd} values between those of the mantle or those of the average crust. Additions of mantle-derived magma to the crust will change the isotopic evolution of the crust; in the absence of any substantial additions to the crust at T_{M1} and T_{M2} the crustal evolution curve will be traced by the purely crustal granitic magmas, which will have ε_{Nd} values that lie within the stippled parallelograms. The *inset at right* shows schematically the distribution of $f_{Sm/Nd}$ values in the crust. (Bennett and DePaolo 1987)

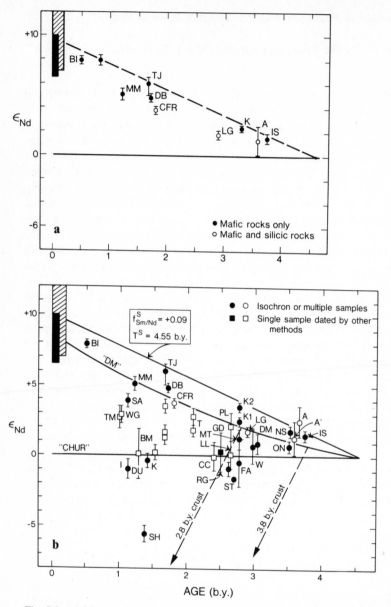

Fig. 5.9. a Initial ε_{Nd} value and age of Precambrian and Early Paleozoic rock suites for which there is little evidence of involvement of older crust in their genesis. **b** Initial ε_{Nd} against age for Precambrian rocks. The fields of young island arc rocks *(solid)* and MORB *(ruled)* are shown at left. *Bl* Bay of Island ophiolite (Jacobsen and Wasserburg 1980); *TM* Town Mountain granite; *DU* Duluth gabbro; *SA* Sierra Ancha diabase; *FA* Fiskenaesset Anorthosite; *PL* Preissac-Lacorne batholith; *LL* Louis Lake granodiorite; *GD* Great Dyke; *A* and *A'* Amitsoq gneiss (two different ages used) (DePaolo and Wasserburg 1976a,b, 1977); *WG* Westport granite; *CC* Chinese granites; *BM* Blue Mountain; *T* Tassandjanet (Allègre and Ben Othman 1980); *MM* Mount

the time of their formation because a large amount of sediment derived from preexisting continents was incorporated as the crust was generated from the mantle.

5.3.2 Mantle Isotopic Evolution and Model Ages

The initial ε_{Nd} data for a variety of continental rocks are summarized in Fig. 5.9a and b. Figure 5.9a shows data for rock suites that appear to be the oldest (within about 0.1 b.y.) in a particular province, and thus should have initial values approximating that of the mantle at the time of their formation. These rocks invariably have positive ε_{Nd} values, and the younger the age the higher the value of ε_{Nd}. The data clearly show that continental crust has been derived from mantle reservoirs that have characteristics similar to those of the sources of modern oceanic basalts. These mantle reservoirs, by virtue of their positive ε_{Nd} values, must have positive values of $f_{Sm/Nd}$. Referring to Fig. 1.1, this implies that the light REEs are depleted relative to the heavy REEs in the mantle.

Figure 5.9b exhibits a more complete data set on old continental igneous rocks, where there has been no attempt to screen out rocks formed at times more recent than the crust-formation age. The initial ε_{Nd} values are bounded on the high side by the curve drawn through the points from Fig. 5.9a. As expected, the initial ε_{Nd} values cover a region extending to substantial negative values, and the lower limit decreases markedly for younger rocks. The data confirm the importance of crustal involvement in continental magmatism, and suggest that such involvement is more common in more recent times (Allègre and Ben Othman 1980).

The observation that the upper mantle typically has positive ε_{Nd} values means that an empirical model age can be constructed using the average "depleted mantle" ε_{Nd} evolution instead of the chondritic evolution (CHUR). This model age, called T_{DM} by DePaolo (1981c) and similarly denoted elsewhere, is shown schematically in Fig. 5.10. The mantle ε_{Nd} evolution curves used by different workers to calculate the model age vary, but most lie between the upper two curves shown on Fig. 5.9b.

Marcy (Ashwal et al. 1986); *K* Kiglapait (DePaolo 1985); *SH* Shabogamo (Zindler et al. 1981); *TJ* Tijeras; *DB* Dubois (Nelson and DePaolo 1984); *CFR* Colorado Front Range (DePaolo 1981c); *MT* Munro Township (Zindler 1982); *RG* Rhodesian greenstones (Hamilton et al. 1977); *ST* Stillwater (DePaolo and Wasserburg 1979a); *K1, K2* Kambalda – Kanowna (McCulloch and Compston 1981); *LG* Lewisian (Hamilton et al. 1979b); *W* Bighorn gneiss (Carlson and Diez de Medina 1982); *ON* Onverwacht (Hamilton et al. 1979a); *NS* Nulliak (Collerson and McCulloch 1982); *IS* Isua (Hamilton et al. 1978)

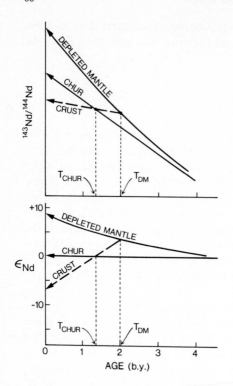

Fig. 5.10. Definition of T_{DM} and T_{CHUR} model ages

5.3.3 Crustal Nd Isotopic Evolution

The evolution of the ε_{Nd} value in crustal rocks after they form from the mantle provides insight into how continents evolve and change with time. In the western United States, extensive mapping and geochronological studies provide the geologic and temporal control necessary to fully evaluate the isotopic data, and make the area an ideal example.

The broad region of Proterozoic crust (Fig. 5.11) can be subdivided into three provinces based on the isotopic data and the geochronological data (Bennett and DePaolo 1987; Farmer and DePaolo 1983, 1984; Nelson and DePaolo 1984, 1985). In area 3 (Fig. 5.12) the oldest rocks are tholeiitic- to calc-alkaline greenstones about 1.7 b.y. old that have initial ε_{Nd} values of $+4$ to $+6.5$. These volcanic rocks are intruded by plutons of gabbro, quartz diorite, and granodiorite that are only slightly younger (Anderson et al. 1971). The plutonic rocks also have initial ε_{Nd} values of $+4$ to $+5$. The ε_{Nd} values strongly suggest that this segment of continental crust was produced from the mantle entirely in an oceanic setting 1.7 b.y. ago. This region, as well as the other areas of the Proterozoic crust, was perforated by undeformed granitic plutons about 1.4 b.y. ago (Silver 1980) and by peraluminous granite in the Mesozoic and Tertiary (cf. Farmer and DePaolo 1983, 1984). The initial ε_{Nd}

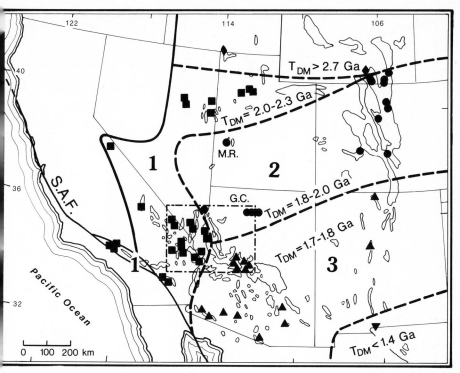

Fig. 5.11. Isotopic provinces of the western US, based on initial ε_{Nd} values in granitic rocks with crystallization ages ranging from ca. 1.8 billion years to Tertiary (Bennett and DePaolo 1987)

values of the rocks, when plotted against crystallization age, define an evolution path for the crust of the area that starts on the mantle evolution curve and progresses linearly to lower values with time. These data show that most of the granitic magmas formed at 1.4 b.y. were produced by melting of the 1.7 b.y.-old crust, and all of the young peraluminous granites were formed by melting of the same type of material.

Province 1 had a similar temporal development, except that the oldest rocks may be slightly older than those of Province 3. The initial ε_{Nd} values of the granitic rocks also define an evolutionary trend (Fig. 5.12), but all of the rocks have lower ε_{Nd} values than the corresponding ones from Province 1. In particular, the ca. 1.8 b.y.-old rocks have initial ε_{Nd} values that are clearly displaced from the mantle evolution curve. These data distinguish the origin and evolution of these two Proterozoic crustal terranes. A geographically intermediate terrane (Province 2) has intermediate characteristics, both in terms of the crystallization ages of the oldest rocks and their ε_{Nd} values. In each province, there was little further addition of mantle-derived material to the crust after the initial episode of crust formation.

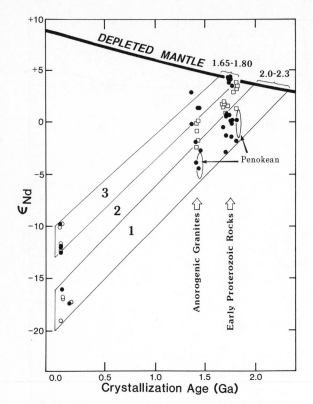

Fig. 5.12. ε_{Nd} versus age for granitic and volcanic rocks of the western US, showing how the geographic groupings (Provinces 1, 2, 3; see Fig. 5.11) also have coherence in their isotopic characteristics (Bennett and DePaolo 1987)

In the area that bordered the Archean craton (Province 1) there was apparently a substantial component of older crust incorporated into the new crust formed in the Proterozoic. In the area farthest from the Archean craton (Province 3) there was apparently no older crust involved. The progression to lower initial ε_{Nd} values outward from the Archean craton in parallel with decreasing crystallization ages suggests that the Proterozoic crust was built onto the preexisting Archean nucleus, and as more and more new crust accreted, there was more effective screening of the Archean contribution from the newly forming crust. This situation is not much different from that observed for the Mesozoic of the western US (Sect. 9.3) except that the scale is larger for the Proterozoic example. It would be consistent with the formation of new crust by successive accretion of magmatic arcs at the edge of the Archean craton. The first arc impinged on the sedimentary apron of the Archean craton and therefore incorporated substantial amounts of older crustal rock. Successive arcs formed rapidly enough that this older component became less and less important. Ultimately, in Province 3, there was negligible involvement of older material. Province 3 appears to be an example of primitive oceanic arcs that evolved to mature arcs with intermediate-composition plutonism in an oceanic area with no involvement of older crust.

Fig. 5.13. ε_{Nd} − age evolution curves for crustal rocks from provinces with model T_{DM} ages ranging from 3.7 to 1.4 Gy (DePaolo 1988). Filled parallelograms indicate fields of existing data

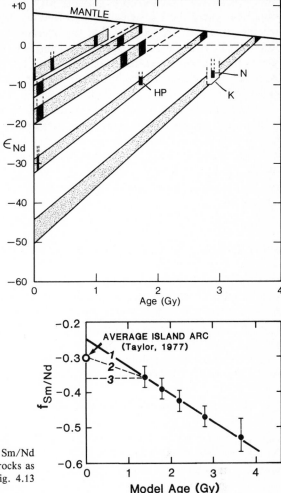

Fig. 5.14. Age dependence of the Sm/Nd fractionation factor for crustal rocks as determined from the data of Fig. 4.13 (DePaolo 1988)

A summary of the ε_{Nd} evolution curves for continental crust of different ages (Fig. 5.13) shows that the slope of the evolution curves appears to be age-dependent. In most cases the data set contains both ε_{Nd} values of granitic rocks of different ages and measured $\varepsilon_{Nd}(O)$ values of the Precambrian basement rocks. The granitic rocks represent samplings of the continental crust at various deep levels and at different times throughout the history of the crust. The agreement between the data from the granitic rocks and the measurements on the Precambrian rocks themselves strongly suggest that this approach gives a good representation of the crustal isotopic composition.

The data show that the preserved remnants of Archean crust have lower Sm/Nd ratios than does crust of Proterozoic age. The derived crustal Sm/Nd ratio appears to increase systematically with decreasing age. This is shown in Fig. 5.14 where the $f_{Sm/Nd}$ value is plotted against model age. It is not clear whether this actually requires that the average Archean crust have been different from younger crust, or that low-Sm/Nd crust (which generally is also low in heat-producing elements) is just more likely to have survived since the Archean. In either case the trend has important implications for the geochemical evolution of the continents.

The trend derived from the Nd isotopic data on granitic rocks is opposite to that derived from sedimentary rare-earth patterns as described by McLennan and Taylor (1984). However, it is possible to reconcile this difference if it is postulated that the Archean sediments are derived from sources with a high proportion of high-Sm/Nd mafic lavas, which is a likely possibility for Archean times, when there was probably abundant continental mafic volcanism.

5.3.4 Resolution of Crustal Ages in Multiply-Metamorphosed Terranes

Because of the relative geochemical coherence of Sm and Nd during metamorphism, the Sm-Nd isotopic data can be used to determine the approximate age of crustal rocks in regions where the information from other geochronological systems has been severely compromised by high-grade metamorphism. A good example is provided by the work of Jacobsen and Wasserburg (1978a), who used Sm-Nd model ages to determine the age of a terrane in Norway where the rocks had been metamorphosed to granulite facies 1.8 b.y. ago.

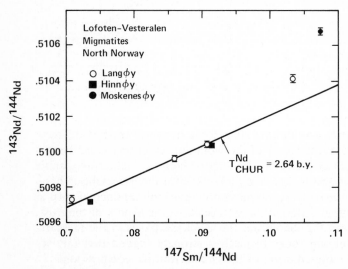

Fig. 5.15. Sm-Nd isotopic data for rocks from Lofoten-Vesteraalen, Norway (Jacobsen and Wasserburg 1978b)

Prior to their study it had been proposed, on the basis of Pb isotopic data (Taylor 1975), that the protoliths of the granulite facies metamorphic rocks were originally formed 3.4 b.y. ago. The Sm-Nd data (Fig. 5.15), indicate that the model crust formation age for these rocks is about 2.6 b.y., and therefore that the interpretation of the Pb data was incorrect. In this terrane, even the Sm-Nd data have been disturbed somewhat by the metamorphism, but it is nevertheless clear that the original age of the rocks is not greater than about 2.7 b.y.

Less conclusive results were obtained on the granulite facies rocks from the Fyfe Hills, Antarctica by DePaolo et al. (1982). The isotopic data for the Rb-Sr, U-Pb, and Sm-Nd systems are shown in Fig. 5.16. The time of the last granulite facies metamorphism had been dated to be about 2.5 b.y. ago, but the age of the protoliths was in question. Pb-Pb data (Sobotovich et al. 1976) had been interpreted as indicating an age of 4.0 b.y. During the granulite facies metamorphism, these rocks had suffered large-scale Rb and U depletion. This is shown by the relatively small slopes of the isotopic evolution lines (Fig. 5.16a and b) after 2.5 b.y. as compared to those necessary to produce the $^{87}Sr/^{86}Sr$ and $^{206}Pb/^{204}Pb$ ratios 2.5 b.y. ago from a mantle ratio. Because of this large change in the Rb/Sr and U/Pb ratios during the metamorphism, the model ages for Rb/Sr and U/Pb are not reliable indicators of the age of the protoliths.

The evolution lines for ε_{Nd} are not changed as much as those for the other isotopic parameters (Fig. 5.16). The model ages fall in the range 3.1 to 3.9 b.y. Because the rocks are in large part metasedimentary, it is possible that more than one, or a range, of crust-formation ages applies. The authors used the detailed aspects of the data to argue that an age of 3.6 b.y. is probably a best estimate for the age of the protoliths, and that Sm and Nd were redistributed to a substantial degree during the metamorphism. In this case the fact that the data can be used to make a qualitative estimate of the protolith age is mainly because Sm-Nd fractionation has not been large; that is, both elements were mobile, but moved together for the most part. Systematic studies of isotopic composition and trace element concentrations in high-grade metamorphic rocks are scarce, so that the detailed behavior of the rare-earth elements during metamorphism is not well understood (cf. Collerson and Fryer 1978).

Another application of the Sm-Nd model age is the work of McCulloch and Wasserburg (1978c) on composite samples of rocks from the Canadian shield. The samples measured were composites prepared so as to reflect the relative areas of exposure of the different rock types in the areas shown in Fig. 5.17a (cf. Shaw et al. 1976). The ε_{Nd} evolution curves (Fig. 5.17b) show that the T_{DM} model ages cluster near 2.7 b.y., except for the Quebec composite, which has an age of 1.4 b.y. In the case of the Churchill province, which is represented by the Saskatchewan composite sample, previous age determinations had clustered at about 1.8 to 1.9 b.y., but the T_{DM} model ages indicate that, in the area of the composite, the crust is mostly made of material that was originally derived from the mantle 2.7 b.y. ago. The concentration of ages

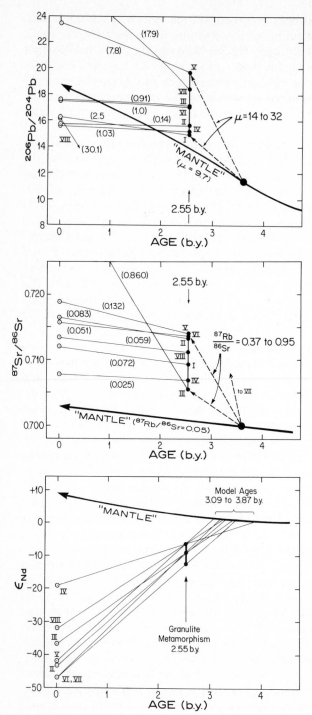

Fig. 5.16. Pb, Sr, and Nd isotopic data from granulite facies rocks from Enderby Land, Antarctica (DePaolo et al. 1982). The model ages of the Rb-Sr and U-Pb systems were severely disturbed in the Late Archean granulite facies metamorphism. The Sm-Nd model ages may also be disturbed, but to a much smaller extent

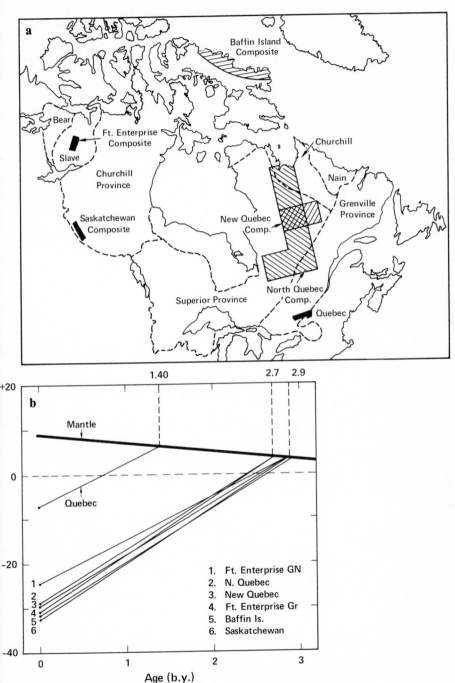

Fig. 5.17. (a) Map of the Canadian Shield from McCulloch and Wasserburg (1978) showing the areas represented by composite rock samples. **(b)** The ε_{Nd} evolution for the Canadian Shield composites

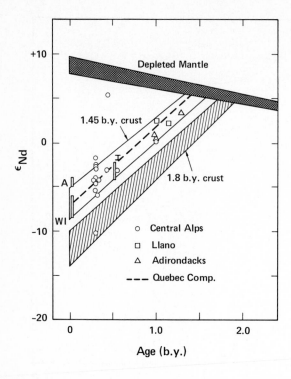

Fig. 5.18. The ε_{Nd} evolution for basement rocks of the central Alps and "Grenville" age rocks from North America. The extrapolation to the mantle evolution curve gives ages of about 1.4 to 1.5 b.y. (Steiger and DePaolo 1986). A: Adamello; I: Ivrea granulites; WI: Western Insubric Line intrusives

at 2.7 b.y. indicates that much of this large region of the North American continent was formed from the mantle at that time.

The basement rocks of the Alps in Switzerland and northern Italy represent a good test of the application of the Sm-Nd model age method. These rocks have been affected by Early and Late Paleozoic orogenic events, as well as the Cenozoic Alpine orogeny. The original age of this crust has been completely obscured by the metamorphisms associated with these events. The Sm-Nd data from granitoids of various ages (Fig. 5.18) indicate that the ε_{Nd} evolution of the crust corresponds to a crust-formation age of about 1.4 b.y. By analogy with rocks in North America with a similar isotopic evolution path (also see Fig. 5.13) it appears that the central Alpine basement is composed of rocks with an origin like those of "Grenville" age from North America. The original crystallization ages were probably about 1.0 to 1.2 b.y., and the older model age represents the effect of an admixture of older crustal rock.

5.4 Sedimentary Rocks

The application of Sm-Nd isotopic studies to sedimentary rocks is based on the fact that clastic sediments are fundamentally just mechanical disintegration products of the igneous, metamorphic, and older sedimentary rocks ex-

Fig. 5.19. ε_{Nd} versus petrographic indicator for sediments from a Tertiary basin in New Mexico (Nelson and DePaolo 1988)

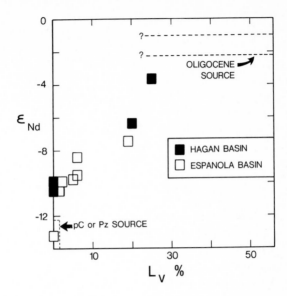

posed in areas above sea level. As discussed above, large regions of the continents can be characterized by consistent and well-defined model ages. When these continental areas are eroded and the comminuted materials transported by streams and deposited, the resulting sediment accumulations generally preserve the model age of the source terranes. Consequently, to the extent that the possible source terranes of a sedimentary formation can be limited, and to the extent that their Sm-Nd model ages are known, the model age measured on the sedimentary rock helps to identify the sediment source area, also referred to as the provenance. This type of information is valuable for reconstructing the history of sedimentary basins.

A test of these assumptions has been carried out by Nelson and DePaolo (1988). They restricted their investigation to small continental basins with only two or three isotopically distinct possible sediment source areas. Provenance determinations based on petrography were also available for the same formations. An example of the results is shown in Fig. 5.19, where the ε_{Nd} value is plotted against a petrographic parameter that is also provenance-sensitive. In general, the isotopic values correlate strongly with the petrographic parameter, but show one noticeable difference. Sediments that were determined on the basis of petrography to be derived from one source only, show isotopic evidence for a significant component from the other possible source. It was concluded that the isotopes were somewhat more sensitive to components present in small amounts.

Overall, the results of Nelson and DePaolo (1988) confirm that Nd isotopic characteristics are a good indicator of sediment provenance, and that the information they provide is complementary to that provided by other methods of provenance determination. They also confirm that interlayered

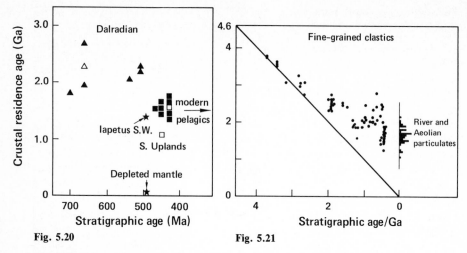

Fig. 5.20. Nd model ages of sedimentary formations that predate and postdate the opening of Iapetus which occurred about 500 million years ago (Miller and O'Nions 1984)

Fig. 5.21. Model age versus depositional age for sediments worldwide (O'Nions 1984)

fine-grained and coarse-grained beds give the same isotopic results. This is particularly important, because it is generally impossible to determine the provenance of fine-grained sediments petrographically.

An application of the Sm-Nd isotope system to provenance determination has been described by Miller and O'Nions (1984) and O'Nions et al. (1983), who measured model ages of Precambrian and Paleozoic sedimentary rocks of the British Isles. They have designated the Sm-Nd model age as t_{CR}, which is similar to the model age designated as T_{DM} (Fig. 5.10). Their data show that the sources of the sediments changed substantially from the time period before the opening of the proto-Atlantic Ocean (Iapetus) to that after the closing. The older sediments generally appear to be derived from source terranes that are Archean in age, whereas the younger sediments appear to be derived from source terranes that are Proterozoic in age (Fig. 5.20). This difference was not identifiable with any other method of provenance determination, because the different age source terranes are composed mostly of similar rock types. The authors suggest that the younger sediments were derived from the Baltic shield, whereas the older sediments were derived from areas of the Superior Province and neighboring provinces now located in northern Canada and Greenland.

O'Nions (1984) and Allègre and Rousseau (1984) applied the method to sedimentary rocks in order to determine something of more global proportions. These workers determined the model ages of sedimentary rocks having depositional ages ranging from early Precambrian to Cenozoic. They assume that the sedimentary rocks, or an average of several sedimentary units from different places on the globe, give an estimate of the average age of the continents at the time of deposition of the sediments. In this way they can trace the average age of the crust through time. In both studies it was concluded that the model age is close to the age of deposition for sediments formed in the Early Precambrian. Sediments formed at more recent times generally have model ages that are substantially greater than the depositional age. The model ages of Late Proterozoic and Phanerozoic age sedimentary rocks seem to reach a constant value of about 1.8 b.y. (Fig. 5.21). The interpretation of this type of data is not unique, but both groups conclude that the continental crust was growing in size during the Early Precambrian but has not grown substantially over the past billion years or so. Another interpretation is that the extent to which sedimentary rocks are derived by erosion of old continental material as opposed to younger orogenic material has changed with time, or that the extent of sedimentary recycling has increased with time. Yet another interpretation is that the extent of recycling of continental crust into the mantle has decreased through time while the mass of the continents has stayed the same.

Taylor et al. (1983) have attempted to determine the average composition of the continental crust by measuring Nd isotopes in wind-blown glacial deposits called loess. They assume that the loess provides a fair representation of exposed crustal rocks over a large enough area that they provide an approximation to the average crust over the whole earth. Although this assumption could be questioned, the resulting model ages (Fig. 5.22) agree well with the model ages derived from measurements of other sedimentary rocks discussed above.

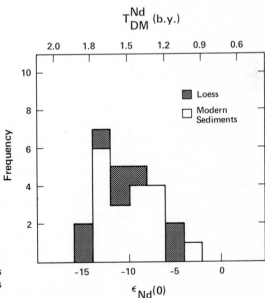

Fig. 5.22. Nd isotopic data on loess and deep-sea sediment samples (Taylor et al. 1983)

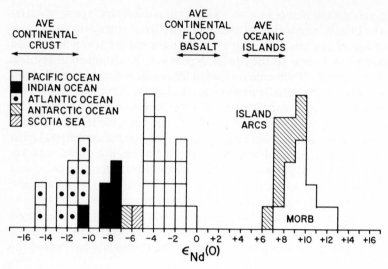

Fig. 5.23. ε_{Nd} histogram of ocean water and Mn nodules (Piepgras et al. 1979)

A number of studies have dealt with establishing the Nd isotopic composition of deep-sea sediments (Fig. 5.22). In general, the sediments of the Pacific Ocean basin have higher values of ε_{Nd} than those from the Atlantic. This observation confirms that the rivers emptying into the Pacific Ocean are draining continental crust that is substantially younger than that drained by the rivers feeding the Atlantic.

5.5 The Oceans

The Nd isotopic composition of water samples from the world's major oceans has been studied most completely by Piepgras et al. (1979). Their results (Fig. 5.23) show that the waters of each major ocean have a fairly restricted range of ε_{Nd} values, but that interocean variability is quite large. The average value of ε_{Nd} in the Pacific Ocean is about -3, whereas the average Atlantic value is about -13. The oceans are thus very poorly mixed with respect to their Nd isotopic compositions. This is a consequence of the fact that the Nd being delivered to the different oceans by rivers differs markedly in isotopic composition, and this Nd is precipitated (i.e., fixed in the sediments) quickly enough that circulation has insufficient time to homogenize the oceans. This is consistent with the estimated residence time of Nd in the oceans, which, at 200 years, is substantially shorter than the estimated global mixing time of the oceans, which is about 1000 years (Broecker 1963).

The implication of these data is that the Nd isotopic composition can be used to fingerprint water derived from different oceans for the purpose of studying ocean circulation. Piepgras and Wasserburg (1980) applied this idea to waters of the Drake Passage, and found that the water mass there, which is generally thought to consist mainly of Pacific Ocean water moving toward the Atlantic, actually contains a strong isotopic imprint of Atlantic ocean water. One problem with the application of this method is that a substantial fraction of the Nd in ocean water is contained in very fine particulate material. Settling of such particulates independent of mixing of the water masses could complicate the interpretation of the data. An attractive aspect of these Nd isotopic studies is that it may be possible by measuring chemical sediments to study paleocirculation patterns in the oceans.

Chapter 6   Correlation of Nd Isotopic Variations with Other Isotopic Variations

With the development of Sm-Nd studies, and shortly thereafter Lu-Hf, it has become increasingly apparent that the relationships between the isotopic variations of different elements are extremely important. This has become even more obvious with the added information provided by the rare gas isotopes ^{3}He, ^{40}Ar, and ^{129}Xe (e.g. Clarke et al. 1969; Craig et al. 1978a, b; Kurz et al. 1982a, b; Lupton 1983; Staudacher and Allègre 1982; Hart et al. 1979). The correspondence between Nd isotopic compositions, and the isotopes of other elements are considered in this chapter.

6.1 Strontium

The initial ε_{Nd} and ^{87}Sr/^{86}Sr values measured in young oceanic basalts from mid-ocean ridges and intraplate islands are correlated, with high ε_{Nd} values corresponding to low ^{87}Sr/^{86}Sr and vice versa (Fig. 6.1). The MOR basalts have the highest ε_{Nd} and lowest ^{87}Sr/^{86}Sr and lie at one extreme of the trend. The ocean island basalts tend to have lower ε_{Nd} and higher ^{87}Sr/^{86}Sr. The slope of the trend indicates that reservoirs with low Rb/Sr have high Sm/Nd ratios, and those with high Rb/Sr have low Sm/Nd ratios. This is broadly consistent with the hypothesis that Nd and Sr isotopic variations in the earth are the result of Sm/Nd and Rb/Sr fractionation caused by magmatic processes that occurred in the mantle in the past (Hawkesworth et al. 1978; DePaolo 1979; Allègre et al. 1979).

The solid line in Fig. 6.1 was drawn through the original small data set (DePaolo and Wasserburg 1976b; O'Nions et al. 1977) as an approximation to the trend. The point $\varepsilon_{Nd} = 0$ corresponds to ^{87}Sr/^{86}Sr $= 0.7045 \pm 5$. Since $\varepsilon_{Nd} = 0$ should represent an unfractionated reservoir, a value of ^{87}Sr/^{86}Sr $= 0.7045$ has been inferred to be associated with this unfractionated mantle reservoir today. From this number, the assumed initial value of ^{87}Sr/^{86}Sr of the earth (BABI $= 0.69898$; Papanastassiou and Wasserburg 1969) and the age of the earth (4.55 b.y.), it can be calculated that the Rb/Sr ratio of undifferentiated mantle material is 0.029 ± 0.003 by weight. The Rb/Sr ratio in primitive mantle cannot be estimated from meteoritic abundances because of the probability of inherent Rb-depletion in the earth (Fig. 3.2). This calculated value is at least four times lower than the average chon-

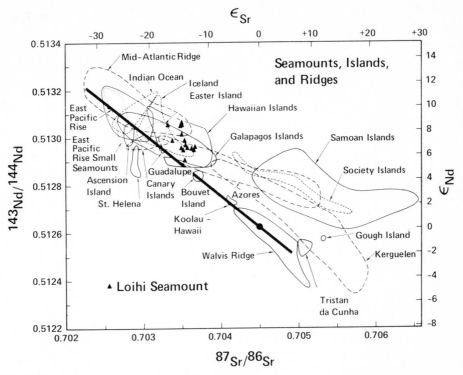

Fig. 6.1. Measured (≡initial) $^{143}Nd/^{144}Nd$ and $^{87}Sr/^{86}Sr$ ratios of young oceanic basalts (After Staudigel et al. 1984)

dritic Rb/Sr ratio (Kay and Hubbard 1978), and agrees with the estimate of 0.033 ± 0.008 made by Hurley (1968), based on other considerations.

The $^{87}Sr/^{86}Sr$ ratio that corresponds to $\varepsilon_{Nd} = 0$ was chosen as the reference value for the definition of the parameter ε_{Sr} (Chap. 3). The resulting ε_{Sr} scale is also shown on Fig. 6.1. Since the original description of the "oceanic mantle correlation trend" for Sr and Nd, many new data have been obtained on volcanic islands, such as the Azores (Hawkesworth et al. 1979a), Samoa, and the Society Islands (White and Hofmann 1982). These data do not lie along the trend but to the right, having generally higher $^{87}Sr/^{86}Sr$ values for the same ε_{Nd}. The divergence of the newer data from the previously proposed trend increases with decreasing ε_{Nd}. A different type of "anomaly" is represented by the data from St. Helena, an island in the South Atlantic. With the exception of St. Helena, the divergent data have been interpreted as evidence of reinjection of crust (altered MORB and continental crust) into the mantle (e.g. Hofmann and White 1982). Since this process should produce isotopic shifts toward higher $^{87}Sr/^{86}Sr$, the original correlation line may still yield a reasonable approximation to the $^{87}Sr/^{86}Sr$ ratio of primitive mantle.

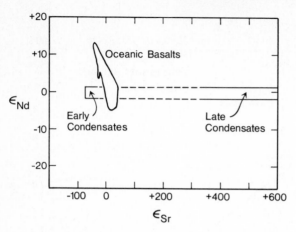

Fig. 6.2. Field of measured ε_{Nd} and ε_{Sr} values of oceanic basalts, representative of mantle reservoirs, and of direct condensates from the solar nebula, which must have been the original material from which the earth accreted

If in error, it is likely to yield an overestimate of the ratio in primitive mantle.

The present-day range of ε_{Nd} and $^{87}Sr/^{86}Sr$ in chondritic meteorites and some achondrites illustrates that meteoritic material has a constant Sm/Nd ratio, but Rb/Sr is highly variable (e.g. Minster et al. 1982). This probably reflects different proportions of high-temperature (low Rb/Sr) and low-temperature (high Rb/Sr) condensates from the solar nebula. The basalt data array intersects the range of compositions of meteorites at a high angle (Fig. 6.2), showing that the isotopic variations in the mantle are not caused by heterogeneous accretion of primitive solar system material. If the earth did accrete heterogeneously, any original Rb/Sr heterogeneities must have been almost completely destroyed, since the width of the basalt data array is small in comparison to the range of $^{87}Sr/^{86}Sr$ values that could be expected in solar nebula condensates.

The $\varepsilon_{Nd} - {^{87}Sr/^{86}Sr}$ correlation can also be treated as a baseline for petrogenetic studies. Significant deviations from the correlation must be an indication of special conditions related to the petrogenesis of the rocks in question. This is illustrated by the divergence of island arc data from the general trend defined by other oceanic lavas (Fig. 6.3). The shift toward higher $^{87}Sr/^{86}Sr$ is probably due to the influence of ocean water in the formation of these rocks. Ocean water contains substantial amounts of Sr with high $^{87}Sr/^{86}Sr$, but since it contains vanishingly small amounts of Nd, there is no effect on ε_{Nd}. The only known material that has elevated $^{87}Sr/^{86}Sr$ in relation to ε_{Nd}, and also has high ε_{Nd} values, is ocean floor basalt that exchanged Sr with heated ocean water circulating through fractures in the oceanic crust at mid-ocean ridges (Chap. 8). The divergence of the island arc data from the main trend has, therefore, been interpreted to be supportive of the hypothesis that the magmas erupted from island arc volcanoes have been generated partly from the melting of ocean floor basalt that is descending into the mantle along subduction zones beneath the volcanoes (Coates 1962; Kay 1980; Gill 1981; Ringwood 1975). The insensitivity of Nd isotopes to hydrothermal alteration

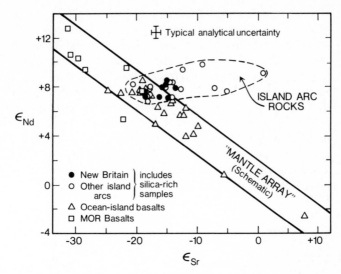

Fig. 6.3. Measured ε_{Nd} and ε_{Sr} values of some island arc lavas showing a trend that is divergent from that defined by other oceanic basalts (DePaolo and Johnson 1979)

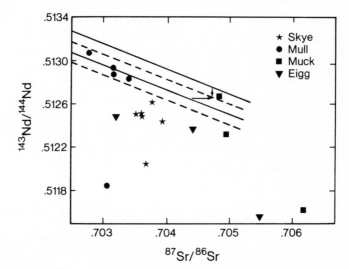

Fig. 6.4. Initial $^{143}Nd/^{144}Nd$ and $^{87}Sr/^{86}Sr$ ratios of Tertiary lavas from the Scottish Hebrides (Carter et al. 1978b). The *solid parallel lines* describe approximately the limits of the "mantle array"

also makes them useful for studying the isotopic composition of older parts of the ocean floor where unaltered basalt has been difficult to find (Jahn et al. 1980; McCulloch et al. 1980; Jacobsen and Wasserburg 1979a).

The correlation line has also been useful for detecting crustal contamination of mantle-derived magma in some cases. A particularly good example is shown in Fig. 6.4. In this case the crustal contaminant has a composition such that its incorporation into mantle magma mainly causes the ε_{Nd} of the magma to be shifted toward lower values, and therefore off of the correlation trend. This would not be discernable with Sr isotopes alone because the contaminant has an $^{87}Sr/^{86}Sr$ ratio within the range of mantle values. Other examples of the use of the correlation line in igneous petrogenesis are discussed in Chaps. 8 – 10.

6.2 Hafnium

The ε_{Hf} values determined on oceanic basalts also correlate with ε_{Nd} and $^{87}Sr/^{86}Sr$ (Fig. 6.5) (Patchett and Tatsumoto 1980a; Patchett 1982; White and Patchett 1984). The relationship between ε_{Hf} and $^{87}Sr/^{86}Sr$ is similar to that between ε_{Nd} and $^{87}Sr/^{86}Sr$. The dispersion in $^{87}Sr/^{86}Sr$ ratios increases with decreasing ε_{Hf}. The correlation between ε_{Nd} and ε_{Hf} exhibits less dispersion. Also, the trend passes very close to the chondritic value for both elements. This observation indicates that the variations of Sm/Nd and Lu/Hf in the earth (both crust and mantle) are quite well correlated, and that the extra dispersion on the ε_{Nd}-$^{87}Sr/^{86}Sr$ and ε_{Hf}-$^{87}Sr/^{86}Sr$ plots does lie in the $^{87}Sr/^{86}Sr$ ratio.

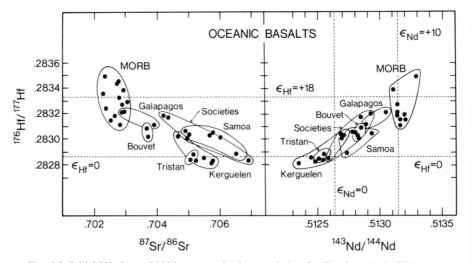

Fig. 6.5. Initial Hf, Sr, and Nd isotope ratios in oceanic basalts (Patchett 1982, 1983)

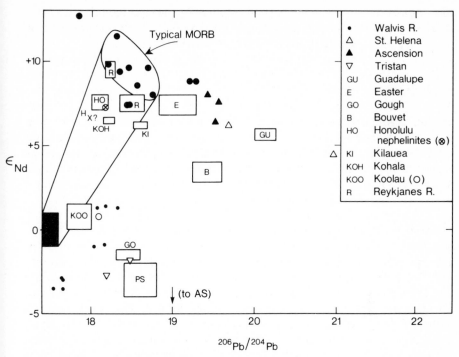

Fig. 6.6. Initial ε_{Nd} and $^{206}Pb/^{204}Pb$ ratios of oceanic basalts

6.3 Lead

The Nd and Pb isotopic compositions of oceanic basalts exhibit some relationship, but not a good correlation (Fig. 6.6). For clarity, only the $^{206}Pb/^{204}Pb$ ratio is plotted here. Mid-ocean ridge basalts have moderately well-correlated ε_{Nd} and $^{206}Pb/^{204}Pb$ (Dupre and Allègre 1980; Cohen et al. 1980), where the highest ε_{Nd} values correspond to the lowest $^{206}Pb/^{204}Pb$ ratios. Some oceanic islands lie on an extension of this trend toward higher $^{206}Pb/^{204}Pb$ ratios (e.g. Ascension, St. Helena, and Guadalupe Islands) (Cohen and O'Nions 1982). Dupre and Allègre (1980) used this trend to estimate that the $^{206}Pb/^{204}Pb$ ratio of primitive mantle is about 21.5 to 22. However, other islands, like Hawaii, Bouvet, and especially Gough and Tristan da Cunha, do not lie on this trend. Also, lavas that have ε_{Nd} and ε_{Hf} values closest to "primitive mantle" values do not have high $^{206}Pb/^{204}Pb$ ratios, but rather low ratios (Koolau basalts). The Hawaiian lavas, in fact, apparently define a trend that is almost normal to the trend used by Dupre and Allègre (1980) to estimate the primitive mantle $^{206}Pb/^{204}Pb$. Because Hawaiian volcanism is associated with primitive He (see below), the "Hawaiian trend" may be more meaningful with respect to indicating the primitive mantle Pb isotopic composition. It is interesting in this regard that the Koolau basalts could have the approximate primitive isotopic values for Nd, Hf, Sr, Pb, and He, and that their Pb isotopic ratios are very close to the "geochron" (Stille et al. 1983).

6.4 Helium and Argon

Data on the relative abundances and isotopic composition of rare gas isotopes add another dimension to studies of mantle evolution. A complete discussion of the basis of these studies will not be attempted here, but some of the results may bear heavily on the problem of the origin of the isotopic correlations. Especially significant are the variations of the ^{3}He/^{4}He ratio in hot springs, gases, and igneous rocks from mid-ocean ridges, oceanic islands, island arcs, and continental geothermal areas (Fig. 6.7). Three components of terrestrial He have been identified: atmospheric helium (^{3}He/^{4}He = $1.384 \times 10^{-6} \equiv R_a$), radiogenic helium ($R/R_a = 0.3$) produced mainly by the decay of U and Th, and "primordial" mantle helium ($R/R_a \geq 30$) Clarke et al. 1969; Craig et al. 1978). The ^{3}He/^{4}He ratio varies systematically in different tectonic settings and correlates to some extent with Sr and Nd isotopic ratios. As summarized by Lupton (1983) and Kyser and Rison (1982), MORBs have fairly uniform values of $R/R_a \approx 9$ and ^{87}Sr/^{86}Sr ≈ 0.7025 (Kurz and Jenkins 1981; Kurz et al. 1982a; Craig and Lupton 1976). Hot spots such as Kilauea volcano, Iceland, Yellowstone, the Ethiopian rift valley, and Loihi seamount have higher ratios ($R/R_a = 15$ to 30 and ^{87}Sr/^{86}Sr $= 0.7030$ to 0.7035) (Kaneoka and Takaoka 1980; Kyser and Rison 1982; Kurz et al. 1982b). Relative to the MORB helium, the helium in these areas apparently comes from a more primitive part

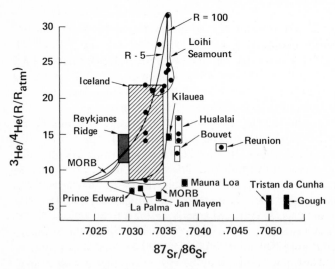

Fig. 6.7. Measured ^{3}He/^{4}He and ^{87}Sr/^{86}Sr ratios of oceanic basalts (Kurz et al. 1982)

of the mantle. Basalts from Gough Island and Tristan de Cunha have lower values of R/R_a (= 6 – 7), but higher $^{87}Sr/^{86}Sr$ (0.704 – 0.705) than the other hot spots (Kurz et al. 1982b). Kurz et al. (1982b) suggest that this component is from a mantle reservoir that has been contaminated with low-R/R_a continental crust. The helium associated with subduction zones generally has $R/R_a = 5 – 8$ (Lupton 1983; Torgersen et al. 1982). This range of values is similar to the Gough and Tristan range, but the subduction zones generally have lower $^{87}Sr/^{86}Sr$ and higher $^{143}Nd/^{144}Nd$.

The He isotope data provide evidence for the reinjection of continental crust into the mantle. They also indicate that hot spots such as Hawaii and Iceland are tapping relatively primitive mantle reservoirs that are high in primordial ^{3}He. The $^{3}He/^{4}He$ data are extremely important because they provide a relatively unambiguous means of differentiating between primitive mantle (high $^{3}He/^{4}He$) and mantle contaminated with recycled crust (low $^{3}He/^{4}He$). For all of the other systems this is difficult because both primitive mantle and contaminated mantle are shifted isotopically in the same direction relative to depleted mantle. The He data suggest that the isotopic correlations are caused by interaction of depleted mantle and primitive mantle (e.g. Hawaii and Iceland) and depleted mantle and contaminated mantle [e.g. Tristan da Cunha, Samoa(?), Azores, others]. Based on this inference, it would appear that the Hawaiian islands data could provide the best indication of the $^{87}Sr/^{86}Sr$ ratio associated with primitive mantle. Analyses of Koolau tholeiites (DePaolo and Wasserburg 1976b; Stille et al. 1983, 1986), which have ε_{Nd} values close to zero, suggest that the mantle $^{87}Sr/^{86}Sr$ ratio is about 0.7042, slightly lower, but within the range of values suggested by DePaolo and Wasserburg (1976b). Because some amount of contaminated mantle may be involved at Hawaii also, this must still be considered an upper limit.

High $^{3}He/^{4}He$ ratios are correlated in a general way with relatively low values of $^{40}Ar/^{36}Ar$ (Kyser and Rison 1982), so the repository of primitive helium (Sr, Nd, and Hf) is also a region of the mantle that was not thoroughly degassed of ^{36}Ar early in earth history. MORBs, on the other hand, have very high $^{40}Ar/^{36}Ar$ ratios (Hart et al. 1979).

6.5 Interpretation of Isotopic Correlations

A unique interpretation of the correlations between Nd, Sr, and Hf isotopes is difficult to construct. It is generally agreed that the isotopic variations observed result from parent/daughter fractionation related to ancient magmatic events (e.g. Hawkesworth et al. 1978; DePaolo 1979; Allègre et al. 1979). However, it is not obvious how to explain the coherence exhibited.

Figure 6.8 shows the systematics of fractionation of reservoirs from a uniform parental reservoir at some time T_f in the past using the Sm-Nd and

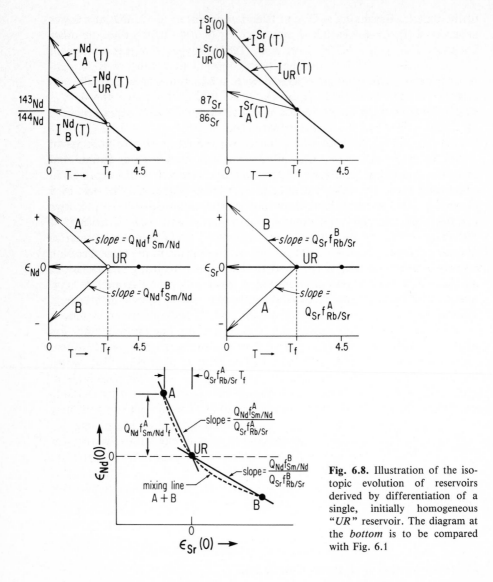

Fig. 6.8. Illustration of the isotopic evolution of reservoirs derived by differentiation of a single, initially homogeneous "*UR*" reservoir. The diagram at the *bottom* is to be compared with Fig. 6.1

Rb-Sr systems for the example. For a partial melt and residual solid derived from a primitive parental reservoir:

Residual solid $f_{Sm/Nd} > 0$, $f_{Rb/Sr} < 0$, $(f_{Lu/Hf} > 0)$

Partial melt $f_{Sm/Nd} < 0$, $f_{Rb/Sr} > 0$, $(f_{Lu/Hf} > 0)$.

The $\varepsilon_{Nd}(0)$ and $\varepsilon_{Sr}(0)$ values (today) in these reservoirs would be simply $\varepsilon_{Nd}(0) = Q_{Nd}f_{Sm/Nd}T_f$ and $\varepsilon_{Sr}(0) = Q_{Sr}f_{Rb/Sr}T_f$, as shown in the lower graph. If line segments are drawn to connect the $\varepsilon_{Nd}(0)$, $\varepsilon_{Sr}(0)$ values of the rock formed from the partial melt and UR or the residual solid and UR, the slopes depend only on the ratios:

$$R_f = f_{Sm/Nd}/f_{Rb/Sr} \; .$$

In order to explain the correlations in terms of these magmatic fractionation events, the ratios R_f must be nearly identical for all melting events, and must be consistent with the predicted fractionation behavior of the elements (Chap. 4). The plausibility of these requirements can be tested only for the Sr-Nd correlation, as distribution coefficients are not determined for Hf. For the Sr-Nd correlation, the required value of R_f is -0.25. The relationship between R_f and the fraction of melting, F, calculated using accepted distribution coefficients for Sm, Nd, Rb, and Sr (e.g. Frey et al. 1978; Irving 1978; Jacobsen and Wasserburg 1979a) and a typical four-phase mantle mineralogy is shown in Fig. 6.9. The required value of R_f obtains for F values of about 0.003 to 0.005. In this range of F, dR_f/dF is large, so small variations of F correspond to relatively large variations of R_f. Only if the value of F is normally very small, and also rigorously controlled during partial melting of the mantle (cf. Stolper et al. 1981; Richter and McKenzie 1984), is it possible to rationalize the correlation as representing simply residual reservoirs from partial melting of primitive mantle.

A modified version of this model would be somewhat more plausible. For example, it is possible that mixing processes in the mantle average out isotopic

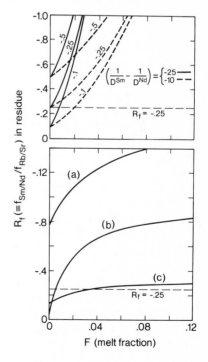

Fig. 6.9. Calculated dependence of the ratio of the Sm/Nd and Rb/Sr fractionation factors of a residual garnet peridotite as a function of the fraction of partial melt removed. The results for two different melting models are shown, fractional melting (top) and batch melting (bottom). A value of $R_f = -0.25$ would produce reservoirs which, after some time has elapsed, could have ε_{Nd} and ε_{Sr} values that would lie along the heavy diagonal line in Fig. 6.1 (DePaolo 1979)

variations so that the slope of the correlation represents only the mean effective value of F, averaged over a large number of partial melt extractions and reduced by continuous mixing with undepleted mantle. This would appear to be consistent with the clustering of values for mid-ocean ridge basalts (Fig. 5.5). It would then have to be surmised that the correlation is caused by mixing a single homogenized residual reservoir with an undepleted or enriched reservoir. This model removes the necessity for a unique F value associated with partial melting, but still requires that the average F value be very small.

Allowable models for the formation of the isotopically distinct enriched and depleted reservoirs span a substantial range. Two possibilities have received recent consideration. One holds that the formation of two (or three) reservoirs happened during an early planetwide differentiation. This type of model has been suggested by Anderson (1982), and would be analogous to current concepts of the early history of the moon (e.g. Taylor 1975, 1978). The other makes the formation of different reservoirs a result of magmatic processes happening over the entire history of the earth. An obvious mechanism for the latter is the formation of a long-lived continental crust, by extraction of partial melts from the upper mantle. The depleted reservoir would be residual material that is constantly mixed by convection. Consequently, it could have a uniform isotopic composition at any time even though its chemical composition could be constantly changing. In both models the correlation would have to be caused by mixing materials derived from enriched and depleted reservoirs or depleted and primitive reservoirs. The second model is an obvious possibility because the continental crust has formed over a long time span and is by far the largest known reservoir that is "enriched", and thus complementary to the depleted reservoir required by the data. The possible involvement of continental crust has an important facet. A likely mechanism for producing mixing between crust and depleted mantle is the injection of sediments into the mantle in subduction zones. Sedimentary material is not necessarily of "average" crustal composition: especially with respect to Rb-Sr characteristics (see Chap. 8). Consequently, a mixing line between subducted sediments and depleted mantle would not necessarily pass through the "bulk earth" composition on any diagram involving $^{87}Sr/^{86}Sr$ (DePaolo 1979; White and Hofmann 1982), Furthermore, because of the dispersion in $^{87}Sr/^{86}Sr$ in continental rocks, a fan-shaped array would be likely, as appears to be the case for ε_{Nd} vs $^{87}Sr/^{86}Sr$. However, in order to preserve the coherent variation of ε_{Nd} and ε_{Hf} (Fig. 5.16), for this model it would be required that the crust and its sedimentary derivatives be homogeneous with respect to Lu/Hf and Sm/Nd (Patchett 1982), which they apparently are not (White and Patchett 1984).

In summary, it appears that the reinjection of continental crust into the mantle may be the dominant cause of the observed isotopic variations in oceanic basalts, but contributions from primitive mantle may also be important in cases like Hawaii and Iceland, where primitive He isotopes are found. It follows from this that the Nd and Hf should be correlated most closely, while Sr and Pb isotopes should show progressively poorer correlations. For Pb isotopes, continental recycling is particularly potent for causing large and unsystematic isotopic variations in the mantle, because of the extreme isotopic heterogeneity of crustal Pb, and the large enrichment of Pb in the crust relative to the mantle. Further discussion of the role of continental recycling is taken up in Chap. 7.

Chapter 7 Models of Crust-Mantle Evolution

The Sm-Nd isotopic data have been used as a basis for testing models of the chemical evolution of the earth which lend themselves to mathematical description. The Sm-Nd system offers a particularly powerful constraint because of the ability to predict the Sm/Nd ratio of the total earth. This characteristic also applies to the Lu-Hf system. The models that have been considered are relatively crude, considering the large number of complex processes that they must attempt to represent. By comparison, for instance, to stellar evolution models, they are simple back-of-an-envelope calculations as opposed to hour-long computer calculations. The geochemical earth models have treated only mass transport, and only in a quasi-equilibrium manner. Various parts of the earth's interior are considered to be "boxes", each generally homogeneous within, and only the transfer of material between these boxes has been considered. There has been little attempt to concurrently model the other aspects of the physical evolution of the reservoirs, such as heat transfer and the fluid dynamic problems associated with the retention of "homogeneity" in the boxes. Instead, there have been attempts to describe the mass transport in sufficiently general terms to allow some relatively robust first-order conclusions to be drawn.

7.1 Material Balance Considerations

The material balance between enriched and depleted reservoirs can be appreciated without mathematical treatment of time-dependent transport. The constraint imposed by the assumption that the bulk earth has $\varepsilon_{Nd} = 0$ for all time, can be written:

$$\sum_j M^j C_{Nd}^j \varepsilon_{Nd}^j = 0 \; , \tag{7.1}$$

where M^j and C_{Nd}^j are the mass and Nd concentration of the jth reservoir. This equation holds at any time during the history of the earth. Similarly, the Sm/Nd enrichment factor is conserved:

$$\sum_j M^j C_{Nd}^j f_{Sm/Nd}^j = 0 \; . \tag{7.2}$$

If the original unfractionated reservoir (CHUR) had a mass, M^0, and Nd concentration, C_{Nd}^0, then:

$$\sum_j M^j = M^0 \tag{7.3}$$

$$\sum_j M^j C_{Nd}^j = M^0 C_{Nd}^0 \ . \tag{7.4}$$

To the extent that all of the reservoirs can be identified within the earth, these equations constrain any model for their formation and evolution.

The essence of the problem can be illustrated with a simple model. An original homogeneous earth, having $f_{Sm/Nd} = 0$ and $\varepsilon_{Nd} = 0$, is differentiated into continental crust (c) and complementary depleted mantle (dm) at some time, T, in the past. The parameters that are known by observation for the earth's crust are the Nd concentration, mass, and $f_{Sm/Nd}$. The age of the crust (i.e., mean age) is also constrained, but will be considered unknown for this calculation. If it is assumed that the earth has an overall chondritic abundance of Nd, and that the core is devoid of Sm and Nd, then C_{Nd}^0 can be estimated to be 1.5 times the chondritic concentration (other estimates, e.g. Anders 1977; are closer to two times chondritic). The parameter known for the depleted mantle is ε_{Nd}^{dm}, based on the ε_{Nd} values of mid-ocean ridge basalts (Fig. 5.5). Recognizing that $\varepsilon_{Nd} = f_{Sm/Nd} Q_{Nd} T$ for the crust, the following relationship is obtained:

$$M^{dm}M^c = (C_{Nd}^c/C_{Nd}^0 - 1) - (C_{Nd}^c/C_{Nd}^0)\frac{Q_{Nd} f_{Sm/Nd} T}{\varepsilon_{Nd}^{dm}} = a + (b/\varepsilon_{Nd}^{dm})T \ . \tag{7.5}$$

The various parameters as estimated by DePaolo (1980) and Jacobsen and Wasserburg (1979b) are given in Table 7.1. The calculated results are shown in Fig. 7.1 for four different values of ε_{Nd}^m today, based on the parameters in column 1 of Table 7.1.

If the average value of ε_{Nd} in MORB $(+10)$ is assumed to be representative of depleted mantle that is complementary to the enriched continental crust, the calculation shows that even if the crust were 4.55 b.y. old − the age of the earth − the depleted mantle could comprise only about 60% of the mass of the total mantle. Existing geochronological data indicate that the actual age of the crust is about 2 b.y. With this further constraint, the conclu-

Table 7.1. Parameters for crust-mantle isotopic evolution models

Parameter	DePaolo (1980)	Jacobsen and Wasserburg (1979b)
M^c	1.8×10^{25} g	2.26×10^{25} g
C_{Nd}^c	22.5 ppm	26 ppm
C_{Nd}^0	0.9 ppm	1.26 ppm
$f_{Sm/Nd}^c$	-0.35	-0.40

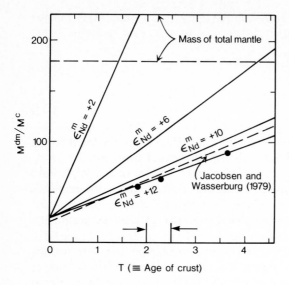

Fig. 7.1. Relationship between the mean age of the crust, the ratio of the masses of the depleted mantle (dm) and the continental crust (c), and the average ε_{Nd} value of the depleted mantle at present. The mass of the total mantle appears as a range on this diagram because of uncertainly in the mass of the continents. The *arrows* at the bottom enclose the range of probable values for the mean age of the crust

sion is that the depleted mantle reservoir comprises only about one-third of the mantle, which would require that two-thirds of the mantle be undepleted, or primitive.

This result appears to be quite robust. It is based on a probable minimum value for C_{Nd}^0 of $1.5 \times$ chondritic. If this were changed to $2 \times$ chondritic (e.g. Jacobsen and Wasserburg 1979b), the calculated M^{dm}/M^c ratios would decrease proportionately. Jacobsen and Wasserburg (1979b) used also larger values for C_{Nd}^c and M^c, and a lower value of $f_{Sm/Nd}^c$. These differences approximately cancel, and calculated M^{dm}/M^c values using their parameters and $\varepsilon_{Nd}^m = +10$ are similar. Furthermore, although this calculation was done for the rather unlikely model of continental crust formation in a single pulse, it will be shown below that the curves in Fig. 7.1 are essentially correct for any plausible crustal growth history if T is regarded as the mean age of the mass of the crust; return of continental crust to the mantle (continental recycling) also has a negligible effect on the calculated curves.

This rather simple calculation implies long-term compartmentalization, or layering, of the mantle (e.g. Wasserburg and DePaolo 1979). However, it is clear from Fig. 5.5 that not all oceanic mantle has $\varepsilon_{Nd} = +10$; values as low as $+4$ are very common, and lower values exist. In the simplest model, as in the above calculation, the mantle is described as consisting of only two parts: one unfractionated ($\varepsilon_{Nd} = 0$), and the other depleted in complement to the crust ($\varepsilon_{Nd} = +10$). If multiple reservoirs are assumed to exist ($\varepsilon_{Nd} = +8, +6, +4$, etc.) by Eq. (7.1) the mass of the reservoir with $\varepsilon_{Nd} = +10$ would have to be even smaller than one-third of the mantle. Since $\varepsilon_{Nd} = +10$ is a common, worldwide number for mid-ocean ridges, any multiple-reservoir model produces equally serious problems. This is considered further in the next section. The only way out of the conclusion that the mantle is semipermanently layered would be if another large enriched reservoir existed in addition to the crust.

7.2 Early Planetary Differentiation

Most discussion of mantle evolution in this volume is distinctly uniformitarian in the sense that the same processes are assumed to have been operating throughout earth history. However, it is quite clear that the period during and immediately after the accretion of the earth must have involved some processes that have not continued to more recent times. The trace element fractionation models (Chap. 4) can be combined with some simplified petrologic models to assess models of planetary-scale differentiation resulting from widespread postaccretion melting, and the degree of compatibility of these with the existing isotopic data. This provides, in addition, an opportunity to examine some of the Nd isotopic data that have been obtained on lunar mare basalts.

A logical place to begin is with a consideration of the early history of the moon. The moon, unlike the earth, has not had much internally generated geologic activity for the past 4 b.y., so more of its early history is still preserved in the rock record. It can serve as a model for early earth evolution. In order to explain the major features of the chemical composition of lunar rocks, it has been hypothesized that the moon underwent widespread melting in its upper 1000 km soon after its formation (cf. Taylor 1978; Walker et al. 1975; Minear and Fletcher 1978; Longhi 1980; Warren and Wasson 1979). This produced, in effect, a global "magma ocean" on the moon, which then crystallized from the bottom to produce a planetary-scale mineralogical and chemical layering. The rare-earth element (Sm/Nd) fractionation that would have been associated with this process can be predicted from the models discussed in Chap. 4. The calculated rare-earth element patterns, assuming that the magma ocean started with chondritic abundances, are in Fig. 7.2a. The patterns represent the crystal accumulates after 31, 53, 60%, etc. of the magma ocean had crystallized. The crystal accumulates represent the lunar mantle from 1000 km depth (0% solidified) to just below the crust (98.3% solidified). The mineralogy assumed is based on the model described by Taylor (1975; also see Drake 1976).

The predicted evolution of ε_{Nd} in the various layers of the lunar mantle is shown in Fig. 7.2b. The curve labeled "KREEP" represents the final residual liquid, and that labeled "Anorthosite" represents the plagioclase crystallizing between $F = 0.4$ and $F = 0.22$, which is presumed to have floated to the top and been incorporated into the lunar crust. The model predicts grossly heterogenous ε_{Nd} in the lunar mantle, which is confirmed by the initial ε_{Nd} values of mare basalts that were derived from the mantle during the period $3.9 - 3.0$ b.y. ago.

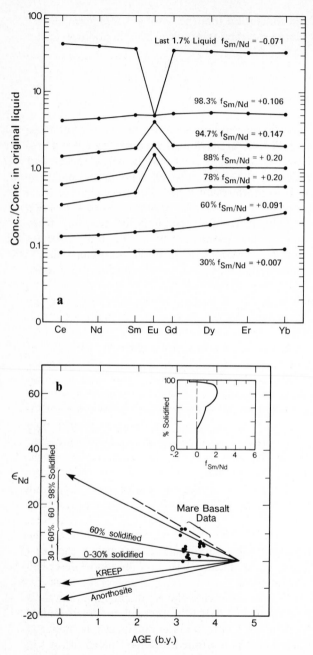

Fig. 7.2. (a) REE patterns for lunar mantle fractional crystallization model; **(b)** ε_{Nd} evolution for lunar mantle and crust and initial ε_{Nd} values for lunar mare basalts (data from Nyquist et al. 1979)

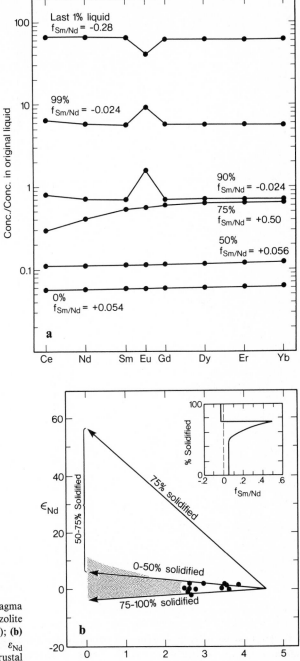

Fig. 7.3. (a) Terrestrial magma ocean REE patterns, lherzolite model (see Hofmeister 1983); **(b)** ε_{Nd} evolution and initial ε_{Nd} values for some Archean crustal rocks (see Fig. 5.9)

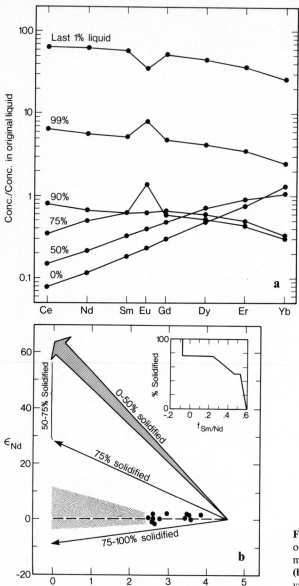

Fig. 7.4. (a) Terrestrial magma ocean REE patterns, picrite model (see Hofmeister 1983); (b) ε_{Nd} evolution and initial ε_{Nd} values for some Archean crustal rocks (see Fig. 5.9)

It has also been suggested that the earth may have undergone a similar "magma ocean" stage during its early evolution (Anderson 1982). For the earth it is more difficult to predict the minerals that would have crystallized because the appropriate pressure range has not been investigated experimentally in great detail. Furthermore, it is not known to what depth the magma ocean would have extended, although Anderson (1982) has suggested 670 km. Hofmeister (1983) used available phase equilibria studies to predict the crystallization sequence for a 120-km-thick terrestrial magma ocean, with an initial composition corresponding to either peridotite or picrite (olivine-rich basalt). The expected rare-earth element patterns for the crystal accumulates formed during crystallization of a terrestrial magma ocean are shown in Figs. 7.3a and 7.4a, assuming that as the crystals form, they trap 5% by weight of the liquid. The ε_{Nd} evolution of the various layers is shown in Figs.7.3b and 7.4b. As for the moon, a large range of ε_{Nd} would be expected to develop in the earth's mantle. However, unlike the moon, the initial ε_{Nd} values of old terrestrial rocks do not show much variation (Fig. 5.9). In fact, for the example of a picritic magma ocean the model predicts that the mantle ε_{Nd} values should lie in regions of the diagram that do not even overlap the data distribution. This probably indicates that any layered structure formed from magma ocean crystallization early in earth history must have been subsequently destroyed.

Although there is no close correspondence between the magma ocean model and the initial ε_{Nd} values of early Archean rocks on the earth, the models may still be of some value. They show that an early terrestrial differentiation of this sort produces mantle reservoirs with very high Sm/Nd ratios that quickly evolve to large positive ε_{Nd} values. Even if this structure was subsequently destroyed by convection, it is possible that vestiges of these early-formed high-Sm/Nd reservoirs lingered until 3.8 or 3.5 b.y. ago, the age of the early Archean metabasalts that have been studied (Sect. 8.4). This might help to reconcile the existence of the relatively high ε_{Nd} values with the apparent lack of preserved pre-Archean continental crust.

7.3 Two-Reservoir Transport Model

A model for continuous crust-mantle interaction is diagrammed in Fig. 7.5. In this model, matter is transferred from the mantle to the crust (crust production) at an arbitrary time-dependent rate $\dot{M}^u$, and from the crust to the mantle (crustal recycling) at a rate $\dot{M}^d$. The lower mantle is of fixed mass and is isolated from the upper mantle and crust. Material transferred to the crust from the mantle is enriched in elements i, j − by factors of D_i, D_j − relative to the concentrations in the mantle at that time. Similarly, material returned to the mantle from the crust has the concentrations $W_iC_i^c(\tau)$, $W_jC_j^c(\tau)$ at any time τ. The upper mantle plus the crust is the system of interest, with mass M^s. The upper mantle is uniform at all times, but the crust need not be uniform. A complete development of the equations for radioactive, radiogenic, and stable isotopes is given by DePaolo (1980), with similar treatment given by Jacobsen and Wasserburg (1979b, 1980b).

The isotopic evolution of Nd in the upper mantle for this model is described by the following differential equation in terms of $\varepsilon_{Nd}(\tau)$, where τ is measured from the time of formation of the earth ($\tau = 0$) toward the present:

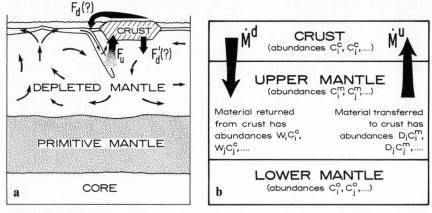

Fig. 7.5. (a) Diagram of layered-earth model with 2-layer mantle and mass transfer between the upper mantle and crust. **(b)** Definition of parameters for the mathematical model (DePaolo 1980)

$$\frac{d\varepsilon_{Nd}^m}{d\tau} = Q_{Nd}\, f_{Sm/Nd}^m \; + \; K_{Nd}(W_{143Nd} - W_{144Nd}) - \lambda_{Nd}^d \varepsilon_{Nd}^m \;, \tag{7.6}$$

where:

$$K_{Nd} = 10^4 \frac{\dot{M}^d\, C_{Nd}^c}{M^m\, C_{Nd}^m}$$

$$\lambda_{Nd}^d = \frac{\dot{M}^d}{M^c}\left(W_{143Nd} + W_{144Nd}\frac{M^c C_{Nd}^c}{M^m C_{Nd}^m}\right) \approx \frac{\dot{M}_d}{M^c}W_{Nd}\left[1 + \frac{M^c C_{Nd}^c}{M^m C_{Nd}^m}\right]$$

$$\equiv \frac{\dot{M}^d}{M^c}W_{Nd}\left[1 + \frac{f_{Sm/Nd}^{dm}}{f_{Sm/Nd}^c}\right] \equiv \frac{\dot{M}^d}{M^c}\left[1 + \frac{X_{Nd}^c}{X_{Nd}^{dm}}\right]\;.$$

Although this equation contains a number of parameters, it is useful for evaluating the implications of any particular history of ε_{Nd} evolution for the upper mantle, which is theoretically measureable (Fig. 5.9a). Equation (7.6) shows the growth rate of ε_{Nd} in the mantle for this model to be the net of the effects due to the mantle Sm/Nd fractionation (1st term) and the recycling.

The simplest model for crustal evolution is one that does not involve any crustal recycling. For this case, the $\varepsilon_{Nd}(\tau)$ evolution of the upper mantle is described by:

$$\varepsilon_{Nd}^m(\tau) \approx Q_{Nd}\int_0^\tau f_{Sm/Nd}^m(t)\,dt \tag{7.7}$$

$$f_{Sm/Nd}^m(\tau) = [M^m(\tau)/M^s]^{D_{Sm}-D_{Nd}-1} = [1 - M^c(\tau)/M^s]^{D_{Sm}-D_{Nd}-1}\;. \tag{7.8}$$

The $f_{Sm/Nd}$ value of the mantle for this model is a simple function of the mass of the crust, and the ε_{Nd} value at time τ reflects the crustal mass history up

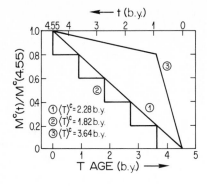

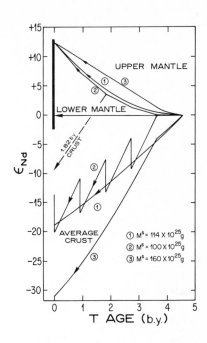

Fig. 7.6. Crustal growth models and calculated ε_{Nd} evolution curves for the mantle and the continental crust. M^s is the mass of the "system", the crust + depleted (upper) mantle, $M^c(t)$ is the mass of the continental crust at time τ; M^c (4.55) is the mass of the crust at present

to that time. Isotopic evolution curves for this model can be easily computed for any crustal growth history. Figure 7.6 shows ε_{Nd} evolution curves for crust and mantle reservoirs, calculated using the parameters in Table 7.1 (column 1) and with the additional constraint that ε_{Nd}^{dm} (4.55 b.y.) = +12, approximately the highest ε_{Nd} measured in an oceanic basalt. Three crustal growth models are shown. An important aspect of the no-recycling model is that the value of $f_{Sm/Nd}^{dm}$ must increase monotonically through earth history, so that the rate of ε_{Nd}^{dm} evolution must be greatest at present. As discussed below, the existing data do not support this model.

The mantle mass-crustal mean age pairs arising from these models are plotted in Fig. 7.1 (dashed line). The gradual growth models, characterized by the mean age of the *mass* of the crust, $\langle T \rangle^c$, are essentially indistinguishable from the single-episode crust formation model as far as the material balance is concerned.

7.4 Sm-Nd and Lu-Hf Data Relevant to the Long-Term Average Rate of Crustal Recycling

The Nd, Sr, Hf, Pb, and He isotopic data from oceanic basalts (Figs. 6.1, 6.5, 6.6, and 6.7) suggest that reinjection of continental crustal rock material into the mantle must be an important process. The apparent evolution of the ε_{Nd}

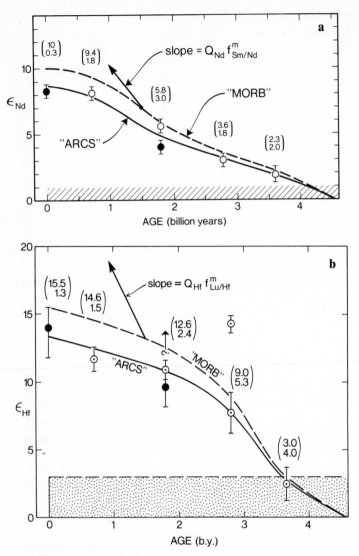

Fig. 7.7. (a) Estimated Nd isotopic evolution for the depleted upper mantle, based on modern island arcs and presumed Precambrian analogs. *Numbers in parentheses* are the ε_{Nd} value and first derivative (in units of b.y.$^{-1}$) of the *dashed curve* at 0, 0.7, 1.8, 2.75, and 3.6 b.y. ago. **(b)** Estimated Hf isotopic evolution (DePaolo 1983b)

and ε_{Hf} values of the mantle through time confirm this, and provide a means of estimating the mass transport rates. The ε_{Nd} and ε_{Hf} values have been determined on recent island arc rocks and on analogous rock suites that are 0.7, 1.8, 2.8, and 3.6 b.y. old (Fig. 7.7 and 7.8). The data indicate that the change with time of ε_{Hf} has been close to zero over the past 2 b.y., and that

Fig. 7.8. Calculated relationship between the continent recycling rate ($\dot{M}^d/M^c$) and the Sm/Nd enrichment factor for the depleted upper mantle ($f_{Sm/Nd}^m$) based on the curve shown in Fig. 7.7a. The *ruled area* is the estimated range of $f_{Sm/Nd}^m$ based on the measured Sm/Nd ratios of mid-ocean ridge basalt

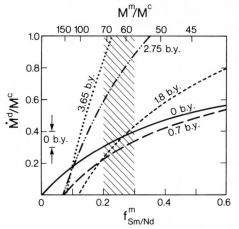

ε_{Nd} has likewise been approximately constant for about the past 0.7 b.y. or more.

The significance of this observation is evident from a simplified form of Eq. (7.6):

$$\dot{\varepsilon}_{Nd}^{dm}(\tau) = Q_{Nd} f_{Sm/Nd}^m(\tau) - \lambda_{Nd}^d \varepsilon_{Nd}^{dm}(\tau) \ ,$$

which is derived by setting $W_{143Nd} = W_{144Nd} = 1$, a reasonable approximation for this case because of the geochemical coherence of Sm and Nd in crustal processes. Because it is known from REE studies of MORB that $f_{Sm/Nd}^{dm}$ must be approximately $+0.25$ or higher (see Fig. 8.1), the observation that $\dot{\varepsilon}_{Nd}^{dm}$ is close to zero requires that λ_{Nd}^d have a substantial magnitude. For Lu-Hf the same argument applies; and $f_{Lu/Hf}^{dm}$ is probably greater than $f_{Sm/Nd}^{dm}$ (Patchett et al. 1981). Recalling the definition of λ_{Nd}^d (see Eq. 7.6), the following relationship can be derived:

$$\dot{M}^d/M^c = \frac{Q_{Nd} f_{Sm/Nd} - \dot{\varepsilon}_{Nd}^{dm}}{\varepsilon_{Nd}^{dm}(1 - f_{Sm/Nd}^{dm}/f_{Sm/Nd}^c)} \ .$$

This equation is graphed in Fig. 7.8 for Nd. The values of $\varepsilon_{Nd}^{dm}(\tau)$, $\varepsilon_{Hf}^{dm}\cdot(\tau)$ $\dot{\varepsilon}_{Nd}^{dm}\cdot(\tau)$, $\dot{\varepsilon}_{Hf}^{dm}(\tau)$ were taken from Fig. 7.7, and it is assumed that $f_{Sm/Nd}^c = -0.37$, and $f_{Lu/Hf}^c = -0.65$. The values used for $\varepsilon_{Nd}^{dm}(+10)$ and $\varepsilon_{Hf}^{dm}(+15)$ are the averages for MORB. The data have been used to estimate the parameters for MORB in the past (dashed evolution curves on Fig. 7.7). For the present time, using a value of $f_{Sm/Nd}$ of $+0.25$, the calculated recycling rate is about $0.35 \, M^c$/b.y. or about $2.5 \, km^3$/year. This corresponds to $f_{Lu/Hf} \approx +0.42$, which is consistent with existing data on Lu/Hf in MORB (Patchett and Tatsumoto 1980a). The same calculation for 0.7 and 1.8 b.y. ago gives approximately the same results for both Sm-Nd and Lu-Hf. This suggests that the mass of the continents has been constant over the past 1.8 b.y., which would be consistent with continental "freeboard" considerations (e.g. Wise 1974).

Using the deduced recycling rate of 0.35 M^c/b.y. and assuming that the crustal mass has been constant since 2.5 b.y. ago, it can be calculated that 42% of the present crust should be older than 2.5 b.y. (42% = $\exp(0.35\tau)$). This is close to the current estimates for the fraction of the present crust that is >2.5 b.y. old. The material balance is not affected by the recycling, so this model still requires that the depleted mantle reservoir represent about 30% of the mass of the entire mantle.

The implication of this model for Pb isotopic evolution is that Pb may be cycled between mantle and crust at an extremely fast rate. From the definition of λ_{Pb}^d:

$$\lambda_{Pb}^d = \frac{\dot{M}^d}{M^c} \left[1 + \frac{X_{Pb}^c}{X_{Pb}^m} \right] ,$$

the estimate for $\dot{M}^d/M^c$, and the estimates for crustal and mantle Pb concentrations, it can be calculated that λ_{Pb}^d may be larger than λ_{Hf}^d and λ_{Nd}^d severalfold. Although the exact value depends critically on the concentrations, which are not known with certainty (compare estimates by Armstrong 1981, and Doe and Zartman 1979), with plausible values one can obtain a residence time for upper mantle Pb of only about 500 million years (residence time $\simeq$ $M^s C_{Pb}^{dm}/\dot{M}^d C_{Pb}^c$). The cycling rate for Sr is probably similar to that of Nd or Hf, but is uncertain because of the complications caused by the sizeable Rb/Sr fractionation that takes place in the crust (including fractionation caused by weathering processes), and because the values of $f_{Rb/Sr}^c$ and ε_{Sr}^{dm} are not sufficiently well known (see Hart and Brooks 1980).

The continent-recycling model fails to account for the primitive He found at sites like Hawaii, Iceland, and the Galapagos (Kyser and Rison 1982; Kurz et al. 1982b). A possible explanation for this would be that plumes from the lower mantle are important at these large "hot spots" (Morgan 1971). Additions to the depleted upper mantle from the lower mantle would serve to moderate the time evolution of ε_{Nd} and ε_{Hf} in the upper mantle. The effect is similar to that of continental recycling, and the transport model can be generalized to include this effect. The resultant differential equation describing the ε_{Nd} evolution is:

$$\dot{\varepsilon}_{Nd}^{dm} = Q_{Nd} f_{Sm/Nd}^{dm} - (\lambda_{Nd}^d + \lambda_{Nd}^p) \varepsilon_{Nd}^{dm} ,$$

where λ_{Nd}^p describes the recharge of the upper mantle by way of additions of primitive lower mantle material:

$$\lambda_{Nd}^p = \frac{\dot{M}^p C_{Nd}^0 f_{Sm/Nd}^{dm}}{M^c C_{Nd}^c f_{Sm/Nd}^c} .$$

The parameter $\dot{M}^p$ is the rate of addition of plume material from the lower mantle to the upper mantle.

Substitution of the estimated parameters for Nd shows that λ_{Nd}^P would be equal in magnitude to λ_{Nd}^d only if $\dot{M}^P \approx 60 \dot{M}^d$. The isotopic effects of "plumes" on the upper mantle are, therefore, expected to be relatively small in comparison to the effects of continental recycling. However, the He data seem to require $\dot{M}^P > 0$, so strictly speaking, the continental recycling rates calculated above must be upper limits. Current rates of magma generation at hot spots could be supported with relatively modest plume fluxes (e.g. $\dot{M}^P < 3\dot{M}^d$). The isotopic effect of such a plume flux is negligible; it requires no significant downward revision of the continent recycling rates. It appears possible to reconcile most of the isotopic data in terms of a model that involves both continental recycling and plumes from the lower mantle.

The models discussed here carry implications for the chemical composition of the crust and mantle. The composition of the upper mantle can be deduced from the composition of the crust and the material balance. Any element in which the crust is strongly enriched should be depleted substantially in the upper mantle. Elements that are not enriched in the crust (such as most of the major elements) will be present in the mantle in concentrations that are close to the primitive value, because the mass of the crust is only about 2% of the upper mantle mass. Depletion factors for some elements with known crustal enrichment factors are shown in Fig. 7.9. Highly incompatible elements like La, Ba, K, Rb, U, and Th should be severely depleted in the upper mantle relative to their original concentrations. This is quite compatible with the chemical composition of mid-ocean ridge basalt (e.g. Kay and Hubbard 1978). A particularly important aspect for geophysical modeling of the mantle is the extreme depletion expected for the heat-producing radioactive elements K, U, and Th. The model predicts that the upper mantle would be devoid of these elements if it were not for recycling of continents. This leaves most of

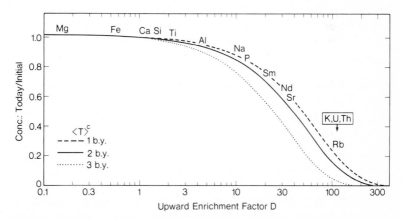

Fig. 7.9. Depletion factor curves for the depleted upper mantle for three different values of the mean age of the crust, $\langle T \rangle^c$. If *K, U,* and *Th* are assumed to behave like *Rb*, then they should be highly depleted in the upper mantle

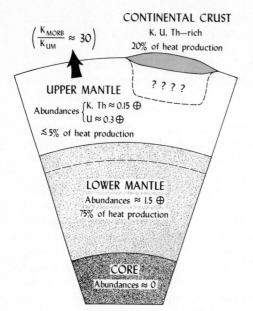

$$\left(\frac{K_{MORB}}{K_{UM}} \approx 30\right)$$

CONTINENTAL CRUST
K, U, Th—rich
20% of heat production

UPPER MANTLE
Abundances $\begin{cases} K, Th \approx 0.15 \oplus \\ U \approx 0.3 \oplus \end{cases}$
≲ 5% of heat production

? ? ? ?

LOWER MANTLE
Abundances ≈ 1.5 $\oplus$
75% of heat production

CORE
Abundances ≈ 0

Fig. 7.10. Radiogenic heat production in the earth based on the depletion factors of Fig. 7.9 and the material balance (Fig. 7.1). According to this model, mid-ocean ridge basalt is enriched by a factor of 30 in potassium relative to the upper mantle

the earth's radiogenic heat production in the lower mantle. The implied present partitioning of radiogenic heat production in the earth is shown in Fig. 7.10.

7.5 Chemical Earth Models

There have been several attempts to use transport models, the systematics of isotopic variations in mantle and crustal rocks, and the framework of plate tectonics and mantle convection, to derive a working model for the chemical evolution of the earth. The intricacies of this task are reviewed in some detail by Allègre (1982). The constraints on these models, or at least the weights of the various constraints, have been perceived quite differently by different workers (Wasserburg and DePaolo 1979; Allègre 1982; Hofmann and White 1982; Zindler and Hart 1986). Nevertheless, there are a set of minimum requirements which must be met for any model to be useful.

1. The material balance derived from Nd (and Hf) isotopes. For all reservoirs identified a useful model must be able to trace their development through time so that the conservation equations are always satisfied.
2. The regional distribution of isotopic values in basalts and other rock types that are representative of the mantle. Models must account for why a certain range of ε_{Nd} ($^{87}Sr/^{86}Sr$, $^{206}Pb/^{204}Pb$, $^{176}Hf/^{177}Hf$, etc.) is common at mid-ocean ridges, a different range is common in island arcs or intraplate islands, and so on. Furthermore, there are correlations between ε_{Nd}, ε_{Sr},

ε_{Hf}, and to some extent, Pb isotope ratios, which yield further constraints.

3. The observed composition of the continental crust, including the age distribution in continental rocks, major and trace element chemistry, and isotopic compositions. The continental crust is the only reservoir that retains direct information about the past.

4. Geophysical constraints on the state of the mantle and its history. The seismic velocity structure of the mantle, its mineralogy, density, temperature, and fusion temperatures provide important constraints on models. Similarly, kinematic models for the mantle, and closely related thermal history calculations, are important.

5. The chemical composition of basalts. Given the isotopic data, the composition of the crust, and the reservoir masses, estimates can be made of the composition of the mantle (e.g. Fig. 7.9), and how it should correspond to isotopic composition. This must be consistent with the chemical composition (especially incompatible elements) of the basalts that are observed to be derived from the mantle. This is a critical problem and is often the departure point between various models. The difficulty arises because the question of compatibility between basalt and inferred mantle composition hinges on understanding magma formation processes and the chemical fractionations they produce. A particular magma formation model (e.g. batch partial melting) may lead to a conclusion of imcompatibility, whereas another model, incorporating, in addition, high-pressure fractional crystallization and wallrock reaction, could lead to a conclusion of compatibility. The relationship between isotopic inferences, and petrologic models based on trace elements and phase equilibria, is extremely close.

6. Plate tectonics, plate velocities, and present and past continent configuration. The time scales for plate movements are different from the reservoir isolation times based on the isotopic data by about one order of magnitude (10^8 years versus 10^9 years). Consequently, any configuration of long-lived reservoirs in the mantle and their expression at the surface as a distribution of ε_{Nd} values in different tectonic settings must be insensitive to the shifting locations of ridges, subduction zones, and continents at the earth's surface.

The "constraints" imposed by these various considerations do not have equal weight, and the exact nature and relative value of each is debatable. In fact, it is extremely difficult to account for all of the possible constraints on models.

An example of an earth model that satisfies at least some of the requirements listed above is shown in Fig. 7.11. The model, modified from that of Wasserburg and DePaolo (1979), is presented here mainly for illustration. It has a long line of progenitors (Morgan 1971; Wilson 1963; Schilling 1973; Sun and Hanson 1975a). The model is essentially one of a two-layered mantle. The lower mantle is relatively primitive material that has remained unfrac-

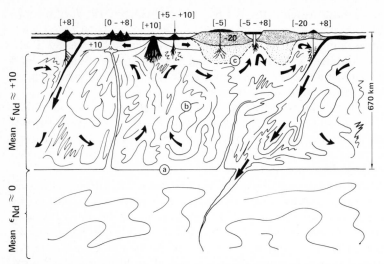

Fig. 7.11. Earth model based on isotopic data from basalts. *Numbers in brackets* at the top are the typical ε_{Nd} values of basalts erupted in the indicated tectonic settings; *from left to right –* oceanic island arcs, intraplate island chains (hot spots), mid-ocean ridges, off-ridge small-volume seamounts, continental cratons, continental rifts, continental margin magmatic arcs. The upper mantle – lower mantle boundary is placed at a depth of 670 km (*a*). The upper mantle contains ribbons of recycled crustal material and material injected from the lower mantle (*b*). The continents are underlain by mantle lithospheric material that is isolated from the convecting mantle (*c*). The lower mantle convects separately but there is some exchange between the upper and lower mantle

tionated with respect to rare earths, Rb, Sr, and most other lithophile elements, but has lost Fe, other siderophile elements, and chalcophile elements to the core, and has lost some, but not all, juvenile volatiles during an early degassing event. The upper mantle, which extends down to 600 – 1 000 km depth, and the crust, make up the upper layer, which has an overall composition similar to that of the lower mantle. The upper mantle is depleted in those elements that are enriched in the continental crust. It is essentially devoid of juvenile volatile elements. The continental crust receives fractionated contributions semicontinously from the upper mantle at magmatic arcs. Material from the crust is reinjected into the upper mantle in subduction zones and stirred back into intimate contact with other mantle material. The upper mantle convects separately from the lower mantle to maintain isolation. Mantle lithospheric keels attached to the continents may also exist. The basic feature of this model is that the upper mantle layers are differentiated, having been previously tapped of volatiles and magmaphile constituents that are stored for long times in the continental crust, the oceans, and atmosphere. The thickness of the layer is based on the Nd material balance.

The surface distribution of ε_{Nd} values, shown in brackets on Fig. 7.11, is explained in the following way. At mid-ocean ridges, large-scale upwelling brings hot upper mantle material to shallow depth where adiabatic decompression causes extensive melting. Magma emplaced into the rifting oceanic crust therefore tends to average over a sufficient volume of the mantle to yield a relatively uniform ε_{Nd} value ($+10$) that is a good representation of average upper mantle.

Intraplate magmatism tends to preferentially sample mantle material that is relatively rich in recycled crustal components. This produces a relatively large range of ε_{Nd} values. The variability of ε_{Nd} in a given region is dictated by the details of the mantle convection pattern. Regions of the mantle that are tapped for magma relatively soon, after receiving recycled crustal rock, will tend to exhibit greater, and less systematic isotopic variations than regions that have been stirred for long times since last receiving recycled crust. An example of a region that may be receiving poorly-homogenized mantle is the South Atlantic, where some of the material subducted at the Peru-Chile trench may be returning quickly to the near-surface region to produce volcanic rocks that are isotopically diverse (e.g. Gough, Tristan da Cunha, Walvis Ridge, Bouvet). In addition, low ε_{Nd} values can be found at large "hot spots", like Hawaii, where isolated "plumes" of material from the lower mantle have risen through the upper mantle. Some continental flood basalt provinces may also be related to plume activity. All magmas erupted through the continents may be affected to some extent by contamination. The continental keel is another reservoir where long-term storage may take place. Its ε_{Nd} value is unspecified; it could be highly heterogeneous like the continents (e.g. Hawkesworth et al. 1984).

This model can satisfy the material balance constraint, and, in addition, encounters no difficulty in tracing the system back through time. The age and composition of the continental crust, and the isotopic composition of new crustal segments through time (Fig. 5.9), are well accounted for by the model. With regard to the geophysical considerations, the important requirements for this model are: (1) a two-layered convection configuration for the mantle, and (2) fairly rapid homogenization in the upper mantle. The former is considered to be possible providing the lower mantle is inherently more dense than the upper parts of the mantle (Richter and Johnson 1974; Olson and Yuen 1982). However, this requires a thermal boundary layer at the upper mantle-lower mantle interface, across which most of the mantle heat flow observed at the earth's surface must be conducted. Spohn and Schubert (1982) consider this to be unlikely if not impossible, whereas Richter and McKenzie (1979) regard it as likely, if not required. The Nd material balance would allow placement of the mantle interface at $650-700$ km depth, which corresponds to the marked seismic velocity discontinuity (Johnson 1967, 1969; Niazi and Anderson 1965), hence the isotopic and seismic data are reasonably consistent with regard to the placement of the major mantle discontinuity if there is two-layer convection. The time and length scales of homogenization in the upper mantle are not well constrained (e.g. Hofmann and Hart 1978). Hofmann and Magaritz (1977) have shown that diffusion is ineffective on the length scales involved, but Richter and Ribe (1979) and Richter et al. (1982) suggest on the basis of simulations of mantle flow that stirring, or "eddy diffusion", in the upper mantle can be very effective. The model is also reasonably consistent with the chemical composition of basalts (Chap. 8) although here a number of degrees of freedom are available in the interpretation of chemical compositions. Finally, the model encounters little difficulty with plate motions.

Armstrong (1981) and Doe and Zartman (1979) have described models that are quite similar in many respects to that outlined here. Variations on the theme have been proposed by Dupre and Allègre (1980), Allègre et al. (1981), Davies (1981), and Hofmann and White (1982). Anderson (1981, 1982) has proposed a quite different model. The construction of better mantle-crust evolution models will probably continue to be a major focus of future research. Aspects of all of the models proposed may need to be incorporated into any comprehensive model. Isotopic data will continue to provide important and increasingly strong constraints on such models, as the base of systematic and precise data increases.

**Part III   Nd Isotopic Variations –
Petrogenetic Studies**

Chapter 8   Oceanic Crust and Mantle

8.1 Oceanic Basalt

The Nd isotopic data provide an important addition to petrogenetic modeling studies of igneous rocks because they provide previously unavailable information about the chemical composition of magma sources. Oceanic basalts present the possibility of applying the isotopic data for this purpose in a relatively straightforward way because there is little likelihood that the magmas are modified by contamination with old continental crust between the time they are generated in the mantle and when they are erupted at the surface. The Nd isotopic ratio (ε_{Nd} value) measured on a lava can therefore be safely assumed to represent the ratio in the mantle where the magma originated. Using an appropriate model the ε_{Nd} value can be used to infer some of the trace element characteristics of the mantle magma source, in particular the Sm/Nd ratio. This can, in turn, be compared to the Sm/Nd ratios of the lavas to evaluate petrogenetic models (Chap. 5). This approach requires assumptions, but is valuable because it allows the internal consistency of the models to be evaluated. The trace element partitioning models provide the basis for evaluating the relationship between the Sm/Nd ratio of a basalt magma and its source in the mantle. A sampling of data for oceanic basalts is shown in Fig. 8.1.

The major assumption that will be made here is that the mantle can be considered as a mechanical mixture of three types of material: depleted peridotite, undepleted or primitive peridotite, and recycled oceanic crust with admixed continental crustal sediment (Fig. 8.1 a). The models of Chapter 7 suggest that the $^{147}Sm/^{144}Nd$ ratio of these materials can be estimated if it is also assumed that:

1. Undepleted peridotite has the chondritic $^{147}Sm/^{144}Nd$ ratio.
2. The T_{CHUR} model age of the depleted mantle is equal to that of the continental crust.
3. The T_{CHUR} model age of the continental crust is the same as that of the average sediment subducted into the mantle.

The first assumption has been discussed previously at some length (Chap. 5). The second assumption follows from the model that the continental crust and the depleted upper mantle are complementary reservoirs. The last assumption is borne out by data on sedimentary rocks. Using data from

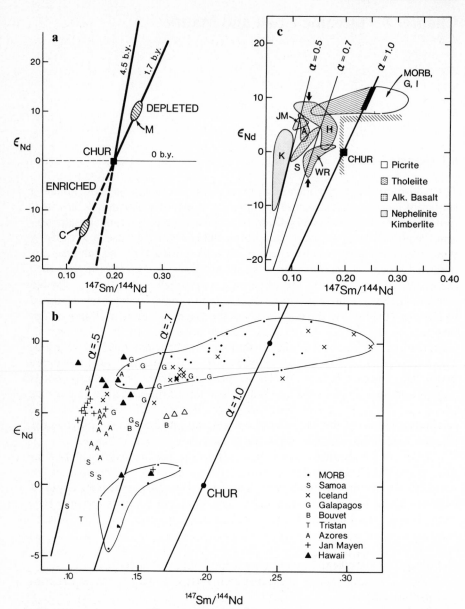

Fig. 8.1. (a) The ε_{Nd} and $^{147}Sm/^{144}Nd$ values estimated for the undepleted mantle (*CHUR*), the depleted upper mantle (M) and the average continental crust (C). A partial melt derived from any one, or any mixture, of these reservoirs will lie to the left depending on the Sm/Nd fractionation associated with partial melting (Fig. 8.2). **(b)** The ε_{Nd} and $^{147}Sm/^{144}Nd$ values of some oceanic basaltic lavas (data from O'Nions et al. 1977; DePaolo, unpublished; Chen and Frey 1983; Hawkesworth et al. 1979a; Stille et al. 1983; White and Hofmann 1979; Cohen et al. 1980; Richardson et al. 1982; Carlson et al. 1978; Zindler et al. 1979). **(c)** Summary diagram showing data on kimberlites (see Fig. 10.9 for data sources)

O'Nions (1984) and Allègre and Rousseau (1984) and an estimated age of the sedimentary mass of the earth, it can be calculated that the T_{CHUR} age of an average sediment is presently very close to 1.7 b.y. Although the sediments presently being deposited have somewhat younger model ages, averaging about 1.2 b.y., it is likely that the average sedimentary material currently in the mantle was subducted ca. 0.5 b.y. ago. The assumption of a 1.7 b.y. T_{CHUR} age for the average recycled component in the mantle thus is reasonable. This age is also very close to the average T_{CHUR} age of the continents (e.g. Nelson and DePaolo 1985).

The resulting ε_{Nd} values and $^{147}Sm/^{144}Nd$ values for the two types of mantle peridotite and average recycled crust are colinear on Fig. 8.1 a. The line drawn between them gives the coordinates of all possible mixtures of the three materials. Partial melting of any mixture of the three produces a magma with a lower $^{147}Sm/^{144}Nd$ ratio, but with the same ε_{Nd} value. The contours of $a = 0.5$, 0.7, and 1.0 represent lines of equal $^{147}Sm/^{144}Nd$ fractionation relative to the mantle material with the same ε_{Nd} value (Chap. 4).

There is a good correlation between the major element characteristics of the rocks and the a value relative to the model mantle reservoirs. The samples that have a values between 1.0 and 0.7 are all classified as tholeiites. Those with a values less than 0.7 are alkali basalt, basanite, and nephelinite. Samples with a values greater than 1.0 are picrites or Mg-rich basalts that have particularly low rare-earth element concentrations. Lavas that would be categorized as transitional (olivine basalts from Hawaii for example) have a values of about 0.6 to 0.7. The relatively few exceptions fall into two categories. Tholeiitic basalts having a values less than 0.7 are those that are quartz-normative, and thus probably owe their low a values to more extensive crystal fractionation. Certain other lavas (two alkali basalts from Kerguelen for example) have a values that are too high. As noted in a later section, this is common for continental basalts, and could be evidence for continental lithospheric material beneath Kerguelen.

The relationship between the a value, the mineralogy of the magma source, and the percentage of melting is summarized in Fig. 8.2. For any given melt fraction, the a value is lower if there is more garnet in the residue. For any given residue mineralogy, the a value is lower if the melt fraction is smaller. From the data of Fig. 8.1 b, it must then be concluded that alkali basalts and nephelinites form by smaller percentages of partial melting than do tholeiitic basalts, and come from mantle source materials that contain more garnet than the sources of tholeiitic basalts. This conclusion is in excellent agreement with those derived from experimental and theoretical phase equilibria studies of the origin of basalt (Green and Ringwood 1967; Ringwood 1975; Green 1971; Mysen and Kushiro 1977; Nicholls et al. 1971), which suggest that alkali basalt (or alkaline picrite) is formed by melting of the mantle at greater depth and in the presence of residual garnet, whereas tholeiitic basalt (or picrite) originates by melting at smaller depth with garnet probably absent from the residue.

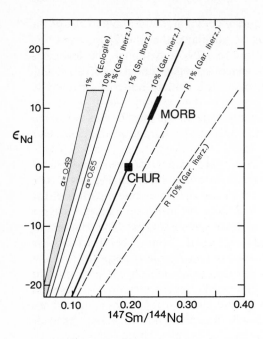

Fig. 8.2. ε_{Nd} versus $^{147}Sm/^{144}Nd$ showing effects of different percentages of partial melting and different garnet contents of the source rock, which lies along the heavy line through MORB and CHUR, on the Sm/Nd fractionation associated with partial melting (see Chap. 4). R denotes residual solid

   The isotopic effects of recycled crustal material in the mantle are also qualitatively predicted by this model. Considering kimberlites as well as the oceanic basalts (Fig. 8.1c) the data show that there is a general correlation between the a value and the ε_{Nd} value. The samples with lower ε_{Nd} values also tend to have lower a values. The lower ε_{Nd} values should be characteristic of parts of the mantle that have a particularly large amount of recycled continental material. These parts of the mantle can also be expected to be richer in garnet for two reasons. First, subducted sediment (or any typical crustal rock) is aluminous, and the addition of aluminum to mantle peridotite increases the garnet content relative to olivine and pyroxenes. Perhaps more importantly, subducted sediment would tend to be associated with subducted ocean floor basalt, which would convert to eclogite (garnet-clinopyroxene rock) at high pressure in the mantle. Therefore, garnet-rich parts of the mantle, which would tend to produce magma with low apparent a values, should also have the lowest ε_{Nd} values.

   The correlation between a and the ε_{Nd} value should not be perfect. One problem is that the ε_{Nd} value of contaminated mantle that contains a particular amount of recycled crustal rock can vary depending on the age of the crustal material. Another problem is that the fraction of melting can vary independently. It appears, for instance, that a values close to 1.0 (large melt percentage) are common at mid-ocean ridge settings but uncommon elsewhere. This results in an area of the diagram, demarcated by the hachured line, where no oceanic basalts plot. Possible exceptions to this are the samples from Oahu and the Walvis Ridge that have ε_{Nd} values close to 0. These near-chondritic

ε_{Nd} values may be indicative of primitive mantle plumes. More extensive melting could reasonably be expected to occur in rising plumes coming from deep in the mantle. Another possibly significant feature is the relatively low a values found in lavas that have ε_{Nd} values of about $+5$ and which are not particularly low in silica. This type is most clearly exemplified by the lavas of Jan Mayen.

A reasonable amount of consistency between the Sm/Nd fractionation, the a value, and the major element composition of basalts is obtained by using the mantle model proposed here. The Nd isotopic data add an important dimension to the study of trace element variations in oceanic basalts, because they give information about the rare-earth pattern of the magma sources. This allows for an improved systematization of the rare-earth element variations. For example, the fact that the Sm/Nd ratio of tholeiitic basalt from the Walvis Ridge ($\varepsilon_{Nd} = -5$, $a = 0.75$) is much lower than the Sm/Nd ratio of tholeiitic basalt from the Reykjanes Ridge ($\varepsilon_{Nd} = +8$, $a = 0.75$) can be understood in terms of the same petrologic fractionation processes (same a value) acting on mantle source materials that had originally greatly different Sm/Nd ratios. Conversely, the similarity between the Sm/Nd ratios of Walvis Ridge tholeiite and Hawaiian nephelinite ($\varepsilon_{Nd} = +7$, $a = 0.55$) is a coincidence. The smaller a associated with the formation of the nephelinite compensated for the higher Sm/Nd ratio of its mantle source. In this case the petrologic evolution of the two magmas was much different even though the final Sm/Nd ratio is the same.

The consistency of the model with regard to the concentrations of Sm, Nd, and the other rare-earth elements is less definite because of the number of unknowns that enter the problem. The most fundamental of these is the rare-earth element concentrations in the undepleted mantle. The simplest assumption is that the earth (including the core) has chondritic concentrations, and therefore that the mantle has about 1.5 times the chondritic concentrations. Other models proposed for the earth suggest slightly higher concentrations (e.g. Ganapathy and Anders 1974).

A nominal rare-earth element distribution pattern can be constructed for the mantle for any given ε_{Nd} value by using the $^{147}Sm/^{144}Nd$ ratio (Fig. 8.1), and the typical shape of moderately light REE-depleted patterns (Fig. 4.4). These patterns are shown in Figs. 8.3 and 8.4. Figure 8.3 gives the pattern associated with an ε_{Nd} value of $+10$, typical of the source of MORB. Some accurate determinations of rare-earth element concentrations in Cretaceous MORB (Jahn et al. 1980) compare well with 5 to 6% partial melts of the model source, assuming that the mineralogy of the magma source is that of a spinel peridotite (Fig. 4.4). However, because these lavas are probably the products of low-pressure fractional crystallization of more primitive magma (e.g. Stolper 1980; Elthon 1979), the primitive magmas were probably a larger fraction partial melt.

Figure 8.4 shows a model rare-earth element pattern for mantle with an ε_{Nd} value of $+6.7$, typical of the sources of oceanic alkali basalts such as

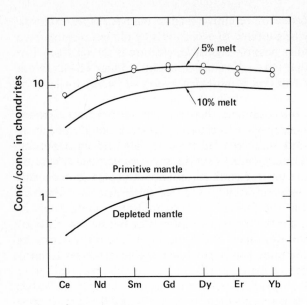

Fig. 8.3. Rare-earth element patterns for primitive mantle, depleted mantle, Cretaceous ocean floor basalt (Jahn et al. 1980), and 5 and 10% partial melt of depleted mantle

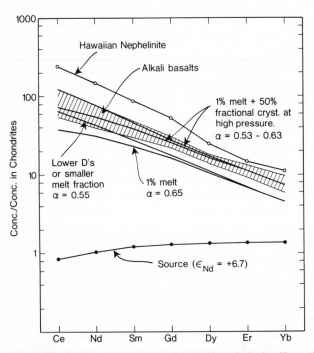

Fig. 8.4. Rare-earth element patterns for "moderately depleted" mantle, typical oceanic alkali basalt, and Hawaiian nephelinite. Model patterns are shown for a 1% partial melt of a garnet peridotite magma source and a 1% partial melt with 50% garnet plus clinopyroxene fractionation (see Chap. 4)

those of Hawaii. This model source material is presumably a mixture of depleted mantle with undepleted mantle or recycled crust. Assuming here that the mantle source has the mineralogy of garnet peridotite, a melt fraction of 1% has a pattern that is parallel to, but below, the field for typical alkali basalt lavas (patterns from Sun and Hanson 1975 b). This model produces the correct amount of Sm/Nd fractionation but underestimates the concentrations significantly. The discrepancy in the concentrations suggests that the simple model of partial melting is insufficient to explain the alkali basalt data. The concentrations could be made to match better if additional processes act on the magma subsequent to its generation by partial melting or if the concentrations in the source material are increased. Intermediate pressure fractional crystallization (Green and Ringwood 1967) or high-pressure fractional crystallization (O'Hara 1968) could improve the agreement between the model and the data. Possible problems with the mineral/liquid partition coefficients (McKay 1986) could also contribute to the discrepancy. The higher elemental concentrations in the nephelinite require smaller partial melt fractions or more high-pressure fractionation or both.

These calculations show that the simplest models do not reproduce the trace element concentrations of alkalic lavas. This general problem has been the subject of numerous investigations (Kay and Gast 1973; Frey et al. 1978; Sun and Hanson 1975 b; Chen and Frey 1983). It appears that the models can come close as long as melt fractions are small (less than 1%). This possibility has been considered unlikely by some workers, but recent modeling studies by Stolper et al. (1981) and Richter and McKenzie (1984) suggest that separation of such small melt fractions is likely under relevant conditions. The alternative that has been proposed is that alkali basalts are produced from mantle that has been enriched in incompatible trace elements by the infiltration of H_2O-CO_2 fluids rich in those elements (e.g. Boettcher and O'Neil 1980; Menzies and Murthy 1980 b; Sun and Hanson 1975; Hawkesworth et al. 1984; Wilshire 1984). This postulated enrichment is constrained by the Nd isotopes to occur just prior to the generation of the magma. This pre-enrichment is not necessary to explain the data if small fractions of melt can separate from the mantle (cf. Sun and Hanson 1975; Frey et al. 1978). Other processes, such as zone refining (Harris 1974) or wallrock reaction involving simultaneous fractional crystallization and assimilation of partial melt from the wallrock (DePaolo 1981 d) may also be important. Advances in understanding also appear to be coming from more sophisticated models of magma generation (O'Hara 1985; Richter 1986).

8.2 Ophiolites

Ophiolites are sections of oceanic crust that have been emplaced onto continental margins (cf. Coleman 1977). They provide an opportunity to study the processes that are involved in the formation of oceanic crust in some detail. From a Nd isotopic viewpoint, they also provide a means to study oceanic crust that is older than Mesozoic, and to assess the isotopic evolution of the oceanic mantle through time.

The Nd-Sr isotopic studies of the Cretaceous Samail ophiolite of Oman (McCulloch et al. 1980) and the Cambro-Ordovician Bay of Islands ophiolite of Newfoundland (Jacobsen and Wasserburg 1979a) produced similar conclusions:

1. The Sr isotopes in the minerals and whole rocks have been seriously disturbed by exchange with hydrothermal fluids derived from ocean water;
2. The Nd isotopes have remained undisturbed, allowing a precise age and a precise initial ε_{Nd} value to be determined for the igneous rocks;
3. The initial ε_{Nd} for both ophiolites is about $+8$; somewhat smaller than the average present-day MORB value of $+10$, and similar to island arc or back arc basin values.

The ε_{Nd}-ε_{Sr} values (Fig. 8.5) for the Bay of Islands ophiolite define a horizontal trend with $\varepsilon_{Nd} = +8$. A two-component mixing curve between basalt and ocean water is shown for comparison. Because of the low Nd concentration in ocean water, extreme water/rock ratios (W/R > 100) would be necessary to significantly affect ε_{Nd}. The largest values of W/R that occur are apparently about 50. This represents a large volume of circulating fluids, but an insufficient amount to affect the ε_{Nd} values. This characteristic of Nd isotopes allows for detailed studies of old oceanic crust that were not possible with Sr isotopes alone. In this case, the mantle array provides a baseline against which the effects of Sr isotope exchange can be measured.

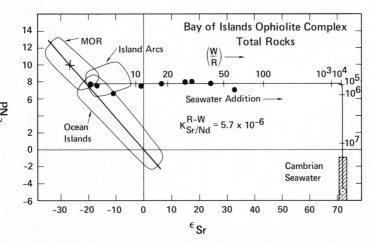

Fig. 8.5. Initial ε_{Nd} and ε_{Sr} values for rocks from the Bay of Islands ophiolite complex (Jacobsen and Wasserburg 1979a). W/R is the water to rock ratio and $K_{Sr/Nd}$ is the ratio of the Sr/Nd ratios in the average rock and seawater

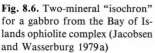

Fig. 8.6. Two-mineral "isochron" for a gabbro from the Bay of Islands ophiolite complex (Jacobsen and Wasserburg 1979 a)

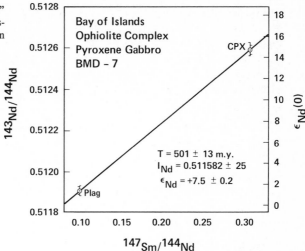

A Sm-Nd age determination made on minerals of a gabbro from the Bay of Islands ophiolite is shown in Fig. 8.6 (Jacobsen and Wasserburg 1979a). The age is determined by plagioclase and pyroxene, the usual case for Sm-Nd isochrons on basaltic rocks (Chap. 2), and is fairly precise. The determined age of about 501 m.y. compares favorably with U-Pb zircon age determinations (Mattinson 1975, 1976) of 504 ± 10 m.y. and 508 ± 5 m.y. made on siliceous intrusives from the same ophiolite, but differs from K-Ar and ^{40}Ar-^{39}Ar dates of about 465 m.y. (Dallmeyer and Williams 1975; Archibald and Farrar 1976). The U-Pb zircon age determinations required some assumptions that were effectively verified by the Sm-Nd ages. This example is a further demonstration of the usefulness of the Sm-Nd method as a geochronometer for mafic rocks where a combination of hydrothermal alteration, tectonic disturbance, and low Rb-Sr ratios make the results from other dating methods unusable or ambiguous. The Ar ages in this case may reflect the emplacement age.

In the Samail ophiolite there has been extensive documentation of shifts in oxygen isotopic composition (Gregory and Taylor 1981) as well as in the ^{87}Sr/^{86}Sr ratio. A composite profile through the ophiolite for Nd, Sr, and O isotopes (Fig. 8.7) shows that the large O and Sr isotopic shifts are mainly restricted to the upper parts of the ophiolite, although deviations of δ^{18}O from the primary magmatic value of $+5.7$ are observed throughout the 7-km section. The implications of this large-scale isotopic exchange for the geometry of hydrothermal convection at mid-ocean ridges has been considered in detail by Gregory and Taylor (1981). A further consequence of the exchange process is the transfer of a substantial amount of ^{87}Sr from the oceans to the oceanic crust. This provides a means to recycle crustal Sr back into the mantle on a significant scale (DePaolo 1980). The fact that this recycl-

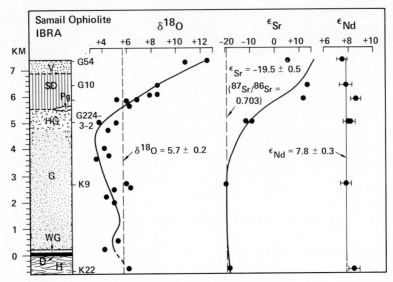

Fig. 8.7. Profiles of ε_{Nd}, ε_{Sr}, and $\delta^{18}O$ in rocks of the Cretaceous Semail ophiolite complex in Oman (Gregory and Taylor 1981; McCulloch et al. 1980)

ing mechanism exists for Sr isotopes, but not for Nd isotopes, is a potentially important consideration in understanding the isotopic variations in the earth's mantle (e.g. Hofmann and White 1982). It is also significant that the ε_{Nd}-ε_{Sr} characteristics of the ophiolites are just those exhibited by some island arc lavas (Fig. 6.3), supporting the notion that some arc lavas are derived from melting oceanic crust, at least in part. These Nd-Sr-O studies of ophiolites provide the best isotopic characterization of oceanic crustal sections that is currently available. The data on the Bay of Islands complex suggest that the mantle source of MORB has not changed significantly over the past 500 million years with regard to its ε_{Nd} value.

8.3 Oceanic Magmatic Arcs

Magmatic arcs built wholly on oceanic crust can be separated on the basis of isotopic characteristics from those built on continental-type crust; the latter type crust being defined as thicker and containing old sialic crustal material, either in the form of crystalline basement or continental margin sedimentary accumulations. Nohda and Wasserburg (1981) have suggested that these two types of arc be referred to as type A and B, respectively.

Most of the data on volcanic arcs built on oceanic crust (type A; Fig. 8.8) fall within a small range of ε_{Nd} from about +6.5 to +10. Most of the points fall very close to the mantle array defined by other oceanic basalts. However, some of the points are displaced to the high-ε_{Sr} side of the mantle array.

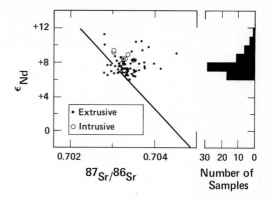

Fig. 8.8. The ε_{Nd} and ε_{Sr} values of oceanic volcanic arcs. Data from DePaolo and Wasserburg (1977), DePaolo and Johnson (1979), Nohda and Wasserburg (1981), Kay et al. (1986), White and Patchett (1984), von Drach et al. (1986)

Those points that fall within the array have isotopic compositions that are identical to a large number of ocean island lavas and just overlap the low-ε_{Nd} limit of the mid-ocean ridge basalt data.

There have been two hypotheses advanced to explain the fact that the ε_{Nd} of the average arc lava is about 2 units lower than that of the average MORB: (1) the mantle source for the arc magmas is different from that of MORB, but the same as that of typical ocean island lavas, or (2) the lower ε_{Nd} results from an admixture of subducted, low-ε_{Nd} deep-sea sediment into the source of the arc lavas (Nohda and Wasserburg 1981; Kay 1980), which would otherwise be the same as the MORB source.

The significance of the shift toward high ε_{Sr} values is also unclear. One possibility is that altered oceanic crust from the subducted slab melted to produce the magmas. Perhaps more probable is the model proposed by Ringwood (1975) or Kay (1980) that a magma or H_2O-rich fluid derived from subducted oceanic crust was added to the overlying mantle wedge, which in turn melted to produce the arc basalts. In either of these cases, an ε_{Sr}-shifted component of the magmas would come from the subducted oceanic crust. However, most of the data that lie distinctly off of the mantle array represent relatively siliceous rocks, dacites and rhyolites (DePaolo and Johnson 1979). These rocks probably are the products of fractional crystallization in shallow magma chambers. Consequently, because the ε_{Sr} shift is best developed in the most differentiated rocks, it is possible that it is mainly caused by assimilation of hydrothermally altered rocks in the volcanic pile itself.

Several of the chemical characteristics of island-arc basalts suggest that complicated models are needed in order to explain their genesis. These chemical traits have been reviewed by Gill (1981) and by Kay (1980). Important aspects of the island-arc basalt chemistry include high abundances of barium, potassium, phosphorus, and some other elements relative to mid-ocean ridge basalts and low abundances of rare-earth elements. Some of these characteristics can be understood in terms of different magma sources, as indicated by the Nd isotopes. For example, higher abundances of barium, potassium, and rubidium would be consistent with the generally slightly lower

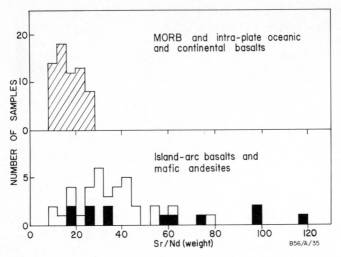

Fig. 8.9. Ratios of Sr/Nd in mafic lavas of oceanic volcanic arcs and comparison with basaltic lavas from other tectonic settings (DePaolo and Johnson 1979)

ε_{Nd} values and slightly higher $^{87}Sr/^{86}Sr$ ratios in the island-arc rocks relative to mid-ocean ridge basalts. On the other hand, this difference is not consistent with the lower rare-earth element concentrations in island-arc basalts. This problem is illustrated by the very high ratios of Sr/Nd (DePaolo and Johnson 1979) in island-arc basalts (Fig. 8.9). The high Sr/Nd ratios are difficult to explain by any simple partial melting or fractional crystallization model. A possible explanation of the difference could be that fluids derived from the subducted oceanic crust are relatively rich in Sr but poor in rare-earth elements, and therefore change the Sr concentration of the overlying mantle wedge without shifting the Nd concentration. However, the solubilities of Sr vs rare earths in fluids at deep mantle pressures are not known. A higher Sr concentration in such fluids would be consistent with the chemistry of other natural fluids, such as ocean water, which is richer in Sr by a factor of about 10^6 relative to Nd. The low rare-earth element concentrations in island-arc basalts could result from the fact that extremely depleted subducted oceanic lithosphere may be the initial source for the magmas, as suggested by Kay (1980).

8.4 Greenstone Belts and Komatiites

The petrogenesis of greenstone belt volcanic rocks is a major problem because they represent the primary occurrence of mafic lavas of Archean age (e.g. Windley 1976). The origin of the ultramafic lavas, called komatiites (Viljoen and Viljoen 1969; Arndt and Nisbet 1982), which occur in greenstone belts, is particularly important because they are high temperature magmas (e.g. Green 1975), and their existence in Archean terranes has led to the hypothesis that the temperature of the upper mantle in Archean time was considerably higher than that of the modern mantle. The Nd isotopic data contribute to this problem because the initial ε_{Nd} values provide information on the rare-earth abundance pattern of the magma sources (e.g. Hamilton et al. 1977).

As noted earlier with regard to mid-ocean ridge basalts, the relationship between ε_{Nd} values and the Sm/Nd values of the magma source is model-dependent. The application of the data to a detailed consideration of the petrogenesis of Archean lavas would be premature because the constraints on the models are still few. Nevertheless, certain aspects of the interpretation can be illustrated. The relationship between the initial ε_{Nd} value of a lava and the Sm/Nd ratio of the magma source is given by Eq. (3.14). The parameter need-ed in this equation is the model age of the magma source (T_s). This is not known, in general, but for purposes of illustration it can be modeled in a way so as to be consistent with what is observed for modern MORB. The following relationship is derived from Eqs. (3.13) and (3.14):

$$f^s_{Sm/Nd} = \frac{\varepsilon_{Nd}(T_x)}{Q_{Nd}\,\varphi\,(4.55 - T_x)} \,,$$

where φ can be evaluated from modern MORB. Using the rule that the Sm/Nd ratio of the magma source should be higher than that of the magma (Fig. 4.4), and that for relatively large percentages of melting there is little fractionation, the highest observed values of Sm/Nd in MORB should be only slightly lower than the Sm/Nd of the MORB magma source. Thus, it is estimated that $f^s_{Sm/Nd}$ is about $+0.25$ for MORB (Fig. 8.1). Using $\varepsilon_{Nd}(0) = +10$ for MORB gives $\varphi = 0.35$. This means that T_s for the MORB magma source is about 35% of the age of the earth ($0.35 \times 4.55 \approx 1.6$ b.y.).

Applying this value of φ to the Precambrian lavas allows some rudimen-tary evaluation of the Sm/Nd ratios of the magma sources. Figure 8.10 gives the Sm/Nd ratios of lavas and calculated magma sources for several suites of Precambrian lavas from greenstone belts. For all of these data there may be some problem with interlaboratory comparisons, especially the data on the Archean lavas, where all of the ε_{Nd} values are small. For the lavas of the Onverwacht series (3.5 b.y. old; Hamilton et al. 1979a) the lavas that are highest in Mg have slightly lower Sm/Nd ratios than that estimated for the magma source using the equation given above. This is consistent with a model of large degrees of partial melting of a peridotite source material that had a rare-earth element pattern close to chondritic, but with light rare earths somewhat depleted relative to heavy rare earths (see Chap. 4). For the Isua greenstones of western Greenland (3.75 b.y. old; Hamilton et al. 1978), a similar conclusion is reached. In this case the higher initial ε_{Nd} value indicates that the source material had a somewhat higher Sm/Nd ratio than that of the Onverwacht basalts, which is consistent with the observation that the Sm/Nd

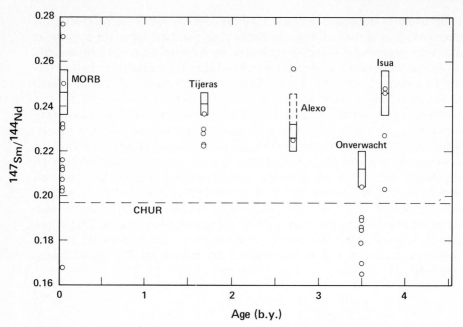

Fig. 8.10. Sm/Nd ratios of lavas (circles) and calculated for magma sources (boxes) for Precambrian greenstone belts and modern mid-ocean ridge basalt (MORB)

ratios of the most mafic lavas are also higher than those of their counterparts in the Onverwacht series. The Tijeras greenstones (1.7 b.y. old; Condie 1980; Nelson and DePaolo 1984), include no komatiites, but they do contain basaltic lavas with high Sm/Nd ratios. The $\varepsilon_{Nd} - Sm/Nd$ relationship for these basalts is also consistent with the models. The Tijeras basalts appear to have been derived from a mantle source that had an Sm/Nd ratio similar to that of modern MORB.

The basalts of the Onverwacht series, Isua, and the Tijeras belt have isotopic and Sm/Nd characteristics that are consistent with simple models of trace element fractionation during partial melting in the mantle, and with the simple model of the relation between the ε_{Nd} values and the Sm/Nd ratio of the magma sources. This is not true, however, for the basalts of the Abitibi belt ("Alexo", 2.7 b.y.) and Cape Smith (1.9 b.y.; Zindler 1982). In both cases the most mafic lavas (komatiites) appear to have Sm/Nd ratios that are higher than those calculated for their respective magma sources (i.e., $\alpha_{Sm/Nd} > 1$). Because the partial melting calculations (Chap. 4) indicate that the magmas must have had lower Sm/Nd ratios than the magma sources, it must be concluded that the source materials had higher Sm/Nd ratios than are indicated by the ε_{Nd} values of the lavas. This is explained as arising because the magma sources were residual material from partial melting that occurred relatively shortly before the formation of the komatiites. The prior partial melting increased the Sm/Nd ratio and decreased the rare-earth concentrations in the mantle magma source (Fig. 4.4), but insufficient time elapsed before the komatiites formed for this to be reflected in their ε_{Nd} values (Hamilton et al. 1977; Zindler 1982). The same type of process must be operating today under Iceland, where picritic lavas have excessively high Sm/Nd ratios (Fig. 8.2).

In general, the ε_{Nd} values and Sm/Nd ratios of Precambrian basalts show patterns that are similar to those of modern tholeiitic basalts from oceanic areas. Most of the Precambrian lavas appear to represent large percentage partial melts (greater than a few percent) of depleted peridotite. In some cases, high-Mg magmas appear to have formed by partial melting of peridotite that had been depleted by magma extraction at an earlier time in the same magmatic event. In other cases the high-Mg magmas appear to be simply large-fraction partial melts. The phenomenon of high-Mg lavas with Sm/Nd ratios in excess of that of the model source material is not common in modern basalts. This could be an indication of differences in the process of formation between greenstone belt-type lavas and those of the modern oceanic crust.

Chapter 9   Continental Magmatic Arcs

9.1 Model of Isotopic Variations in Igneous Source Materials

In this chapter the Nd isotopic data on continental igneous rocks are examined with emphasis on the implications for petrogenesis. The interpretations are based mainly on two concepts. One is a general model of isotopic variations of Nd and Sr in the crust and mantle, and the other is the systematics of mixing between the products of the different magma sources in the crust and mantle.

The model for ε_{Nd} and ε_{Sr} variations (Fig. 9.1) was first described by DePaolo and Wasserburg (1979c), but in a manner that differs in an important way from that given here. They assumed that new continental crust had an ε_{Nd} value of zero when first formed from the mantle, However, subsequent work showed that new continental crust generally had positive ε_{Nd} values initially because it is derived from depleted, rather than undepleted, mantle reservoirs (DePaolo 1980, 1981c; Hamilton et al. 1978). The derivation of Fig. 9.1a is shown in Fig. 9.1b, where the time evolution is explicitly shown. According to the model, the present-day ε_{Nd} value in average crust is a function only of the age of the crust. This follows from the observation that the $f_{Sm/Nd}$ values of crustal rocks fall in a narrow range when averaged over volumes the size of typical igneous source areas (e.g. Haskin and Paster 1979; Ben Othman et al. 1984). For Sr isotopes, however, the crustal ratios are more dependent on rock type or metamorphic history than they are on age. In a crude way the crust is layered, having a granulite facies or more mafic lower part with low Rb/Sr ratios and an upper part that has high Rb/Sr ratios (Heier 1965; Moorbath et al. 1969; Zartman and Wasserburg 1969).

When these semi-empirical models of crustal isotopic evolution are combined with the mantle isotopic variations (Fig. 6.1), the pattern of Fig. 9.1a results. The diagram as shown applies strictly only to the present time, but similar representations could be constructed for other times with appropriate shifts in the ε values and ages shown. The age contours refer to crust-formation ages (equivalent to T_{DM} ages). They do not necessarily represent the crystallization ages for the rocks. For sedimentary rocks the ages would correspond to the mean (Nd-weighted) provenance age, and would normally be different from the age of deposition (O'Nions et al. 1983; Allègre and Rousseau 1984).

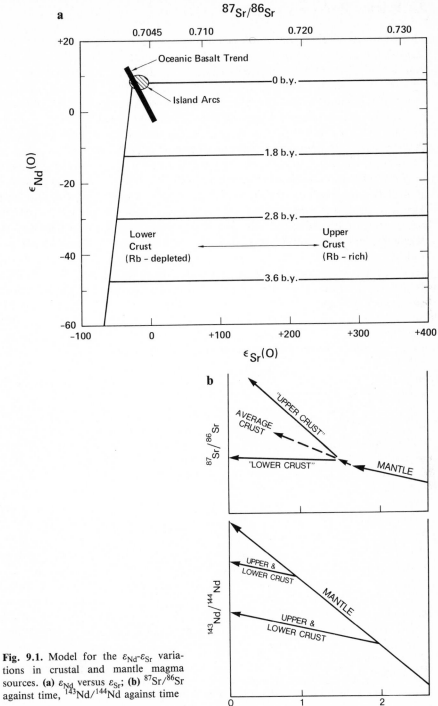

Fig. 9.1. Model for the ε_{Nd}-ε_{Sr} variations in crustal and mantle magma sources. **(a)** ε_{Nd} versus ε_{Sr}; **(b)** $^{87}Sr/^{86}Sr$ against time, $^{143}Nd/^{144}Nd$ against time

9.2 Volcanic Rocks

The relatively limited range of ε_{Nd} values in most of the primitive arcs reflects the limited range in the available magma source materials, oceanic mantle and crust. Only subducted sediments would be different, and they are apparently small in volume. At continental margins there is a much greater range of ε_{Nd} values in the available source materials because of the presence of continental basement rocks composed of old (usually Precambrian) granitic rocks, or thick sections of sedimentary rocks derived from them. One might expect that continental margin magmatic arcs should exhibit some of the isotopic characteristics of primitive arcs because magmas originating from mantle depths should initially have characteristics similar to those of primitive arcs. If the magmas pass through the crust with no interaction they will look like primitive arc magmas. If the magmas interact with the crust, by assimilating it or inducing crustal melting, then the characteristics of the magmas erupted or emplaced at shallow depths will be modified. Their ε_{Nd} and ε_{Sr} values will look like mixtures (modified by differentiation) of primitive arc material and the crustal rocks through which these primitive magmas had to pass. For the most part, this is exactly what is observed in continental margin arcs.

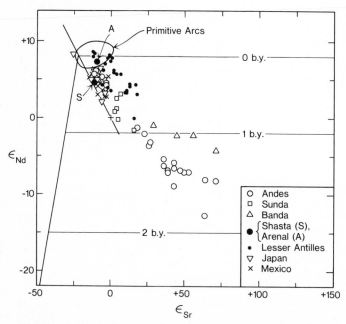

Fig. 9.2. ε_{Nd}-ε_{Sr} variations in continental margin volcanic arcs. Data from Hawkesworth et al. (1979b), Carmichael and DePaolo (unpublished), Nohda and Wasserburg (1981), Hawkesworth et al. (1979c), Whitford et al. (1981), DePaolo (1978a,b)

The Nd and Sr isotopic data on volcanic rocks from continental margin arcs (type B; Fig. 9.2) have several noteworthy aspects:

1. Almost all of the data plot to the right of the line fit through the mantle array, and some of the points lie well to the right by as much as 30 units of ε_{Sr}. This suggests, as does the data from primitive arcs, that hydrothermally altered, ε_{Sr}-shifted oceanic crust is a source for the magma in some of these arcs. The Lesser Antilles lavas have the most pronounced ε_{Sr} shift (Hawkesworth et al. 1979c; Hawkesworth and Powell 1980) of the rocks that have positive ε_{Nd} values. With lower ε_{Nd} the shifts can be explained by contamination with high ε_{Sr} continental rock material.

2. The data array trends away from the primitive arc field well into the field of isotopic compositions of Precambrian crustal rocks. This clearly indicates that crustal contamination and intracrustal melting are important processes in continental margin arcs. The extent of the ε_{Nd} shift due to contamination is dependent on the age of the crust as well as the proportion of material assimilated.

3. There is a concentration of data points near the mantle array with ε_{Nd} and ε_{Sr} values within the range of mantle-derived basalts from other tectonic settings (Fig. 6.1). However, 90% of the data lie outside of the primitive arc field, displaced in the direction of continental rocks. In particular, the main cluster of data in the range $+2 < \varepsilon_{Nd} < 7$ have ε_{Sr} values within the range of primitive arcs, differing from the primitive arcs mainly with respect to ε_{Nd}. Prior to the acquisition of the Nd isotopic data, the Sr isotopic data had been used to argue that primitive arcs and continental arcs were isotopically similar, and therefore, that crustal contamination was not important in the petrogenesis of continental margin arc magmas (e.g. Dickinson 1970). The Nd isotopic data show that this is not the case, even for rocks with $^{87}Sr/^{86}Sr$ less than 0.705.

4. None of the data fall in the negative-ε_{Nd}, low-ε_{Sr} area where granulite facies lower crust would lie. Apparently, such low-ε_{Sr} material does not generally occur in the regions where magmatic arcs develop.

Although a relatively small number of data must be used to represent a large geologic province, a good illustration of the petrogenetic processes responsible for the isotopic patterns is provided by the central Andes and Ecuador (Hawkesworth et al. 1979b, 1982a; Hawkesworth 1982; Francis et al. 1977, 1980; James 1982). The ε_{Nd} values from the Andean lavas vary from $+6$ to -13, spanning almost the entire range observed worldwide. The basement in the central Andes is not very well characterized, but there are indications of ages up to at least 2 b.y. (James et al. 1976). The ε_{Nd} values of the lavas are between the primitive arc values and those expected for the basement. The rocks with the highest ε_{Nd} values are those from Ecuador, which is the region with the least thick continental crust. Those rocks also have other characteristics that indicate minimum interaction with the crust, such as high Sr concentrations and Sr/Nd ratios and low SiO_2 contents (Hawkesworth et

al. 1982a). The data are consistent with a petrogenetic model involving derivation of parental basaltic magmas from the mantle, followed by interaction
with the lower crust and, to a lesser extent, the upper crust. The parental
basalts could have been isotopically identical to primitive arc magmas. Probably all of the magmas, including the Ecuadorian andesites, assimilated crustal
material. Quantitative estimates of the amounts of crustal contaminant are
precluded by the lack of data on the isotopic compositions of the crustal
rocks. If the crust had ε_{Nd} as low as -25, then some of the lavas would have
to contain as much as 50% crustal Nd, which would correspond approximate-
ly to about 25% assimilated crust by mass.

An interesting aspect of the data is that the ε_{Nd} values from a given region
show relatively little variability (3 to 4 units of ε_{Nd}) in comparison to the total
range observed (more than 20 units of ε_{Nd}). The implication is that the
amount of assimilation is controlled in each area by ambient conditions, such
as the local crustal thickness and temperature profile. Another example of this
is provided by the Mesozoic and Tertiary granitic rocks of the western US
discussed in the next section.

The data from the Lesser Antilles (Hawkesworth and Powell 1980;
Hawkesworth et al. 1979c) are considerably different from those of the
Andes. All of the ε_{Nd} values are positive, and all of the data lie particularly
far to the right of the mantle array (Fig. 9.3). Hawkesworth and Powell (1980)
have discussed a number of possible petrogenetic models. The ε_{Sr} shift suggests that some magma (or some components of magma) are derived from
oceanic crust, but the ε_{Nd} shift toward lower values appears to require also
old continental material. The limited data suggest that lower ε_{Nd} values are
observed at the southern end of the arc rather than at the northern end,
although high values are also present throughout the arc. Hawkesworth et al.
(1979c) did not discuss any possible relationship between these data and the
regional geology, but suggested that subducted sediments could be involved.

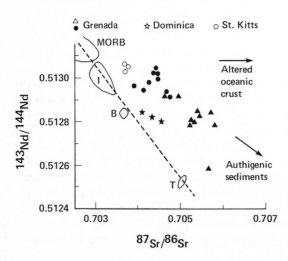

Fig. 9.3. Lesser Antilles Sr and Nd
isotopic data (Hawkesworth et al.
1979c). I: Iceland; B: Bouvet; T:
Triston da Cunha

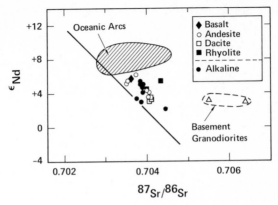

Fig. 9.4. Sr and Nd isotopic data from lavas of the Mexican volcanic belt (Carmichael and DePaolo, unpublished)

The data from the Mexican volcanic belt (Carmichael and DePaolo, unpublished data) show a much smaller range of ε_{Nd} (Fig. 9.4). This is because the underlying basement is too young (Mesozoic and Paleozoic) to have very low ε_{Nd} values. As an example, one sample of granodiorite from the basement has an ε_{Nd} value of $+3$. Consequently, the type of isotopic variations displayed by the Andean lavas is telescoped down to only a few ε_{Nd} units. The relative lack of isotopic variability should not be construed as meaning that crustal interaction was unimportant. In fact, many of the detailed features of the data indicate that it was important. In general, the importance of crustal interaction can only be assessed if the isotopic composition of the basement is reasonably well constrained, and if the basement differs isotopically from the mantle (see Sect. 9.4).

9.3 Plutonic Rocks

The data set for continental margin arcs is sufficiently large (DePaolo 1981 b; Farmer and DePaolo 1983, 1984; McCulloch and Chappell 1982; Liew and McCulloch 1985) that the data can best be discussed case by case. In general, the plutonic rocks have characteristics that are similar to those of the volcanic rocks, but because the areas where plutonic rocks are exposed are also areas where the basement can be examined in some detail, it is possible to better constrain the properties of the potential magma sources.

An example is provided by the Mesozoic and Tertiary plutonic rocks of the Sierra Nevada batholith and its inland extensions in the northern part of the Great Basin of the western United States (Farmer and DePaolo 1983, 1984; Bateman et al. 1963; Kistler and Peterman 1973). A schematic pre-Mesozoic east-west geologic cross-section of the region extending from northern California to north-central Utah (Fig. 9.5a) gives a picture of the probable crustal structure and rock types available in the crust. The western half of this region was underlain by pelagic clastic sediments, mainly of Late Proterozoic

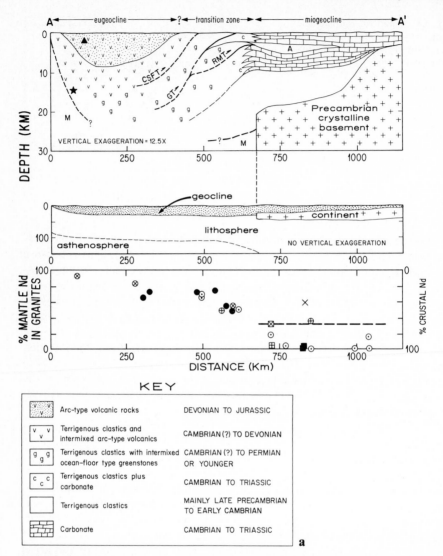

Fig. 9.5. (a) Geologic cross-section of the northern Great Basin, western United States prior to formation of the Mesozoic batholiths (Farmer and DePaolo 1983); (b) ε_{Nd} cross-section; (c) ε_{Sr} cross-section

and Paleozoic age, with intercalated and tectonically interleaved ocean floor basalts and arc-type volcanic rocks. Eastward there is a transition to a shelf-type environment where the sedimentary rocks overlie Precambrian crystalline basement, and are composed both of older clastic rocks and younger carbonates. In the extreme east the sedimentary environment is one of a continental platform.

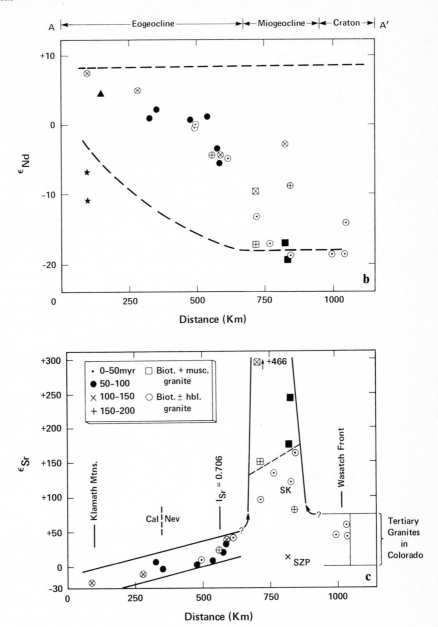

Fig. 9.5 b, c

From the continental margin inland the ε_{Nd} and ε_{Sr} values of the granitic rocks change in a regular fashion from values similar to those of primitive arcs to values that are progressively displaced toward those of old Precambrian crustal rocks. Within this general shift toward crustal values, there are two discontinuities (Fig. 9.5b). One occurs in central Nevada, and the other in western Utah. At the discontinuity in central Nevada, ε_{Nd} values change from a value of about -6 on the west side of values of -18 on the east over a distance of less than 100 km. East of central Nevada, the values are mostly in the range -14 to -18. The second discontinuity is evident only in the ε_{Sr} values. Between the two discontinuities, the granitic rocks have very high and variable ε_{Sr} values (Fig. 9.5c). East of the discontinuity, the ε_{Sr} values are moderately low.

The discontinuity in central Nevada is interpreted as the edge of the Precambrian basement, the locus of an intracontinental rift that formed in the Late Precambrian and originally created the continental margin. The eastern discontinuity marks a change in the crustal magma source characteristics. In the region between the discontinuities, where the Precambrian basement is overlain by a thick section of sedimentary rocks, the basement is apparently relatively rich in Rb and shows no evidence of having a granulite facies lower part. The region east of the discontinuity has only a thin veneer of sediment and apparently does have a low-Rb/Sr lower crust.

When the ε_{Nd} and ε_{Sr} data are plotted together (Fig. 9.6), it is evident that the rocks from the Sierra Nevada and those from west of the edge of the Precambrian basement in northern Nevada have isotopic characteristics that are similar to the continental margin volcanic rocks. All of the data plot to the right of the mantle array except for the granites in the cratonal area. The

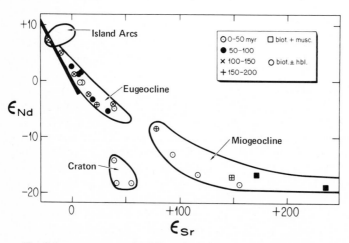

Fig. 9.6. ε_{Nd} versus ε_{Sr} for Mesozoic and Tertiary granitic rocks of the northern Great Basin (Farmer and DePaolo 1983)

"eugeoclinal" and Sierran granitoids are interpreted as resulting from assimilation of various amounts of crustal rock by magmas that originated in the mantle with primitive arc-type isotopic compositions. In northern Nevada the amount of assimilation varies in a regular manner from west to east across the region that was not underlain by Precambrian basement. In the regions underlain by Precambrian basement, most of the granitoids have ε_{Nd} values within a fairly narrow range (-14 to -19), strongly suggesting that they are entirely derived by melting of the crust.

These isotopic patterns can be related to two parameters. One is crustal thickness. For example, the difference in the isotopic compositions of the magmas formed east and west of the edge of the Precambrian basement, is probably related to the greater crustal thickness in the areas underlain by Precambrian basement. However, by comparison with the Sierra Nevada and Peninsular Ranges batholiths, which are partly underlain by Precambrian basement, it also appears that the amount of magma entering the crust from the mantle is an important parameter. In the main magmatic arc relatively close to the subduction zone, which corresponds to the present location of the batholiths, a large amount of mantle-derived magma was probably entering the crust from the mantle. In this situation, the magma that finally rises into the upper parts of the crust is invariably a mixture of mantle and crustal components, even in areas of Precambrian basement. This conclusion can be reached because of the constraints that are available on the isotopic composition of the mantle and crustal endmembers. To the east from the main batholith belts, the amount of magma coming from the mantle was probably much lower, but the temperature of the crust and mantle was elevated sufficiently so that the lower parts of the crust could melt. This is consistent not only with the greater proportion of purely crustal magmas observed in the east, but also with the much smaller volume of intrusive rock that is presently exposed.

A better appreciation for the petrogenetic implications of these data can be gained by considering the isotopic variations on an even larger geographic scale (Fig. 9.7a). The data from the northern Great Basin, along with data from the southern Sierra Nevada, the Peninsular Ranges, and southern Arizona, can be contoured with regard to the lowest ε_{Nd} values of the granitoids in each area, and compared with the currently understood isotopic compositions of the Precambrian crystalline basement rocks. The basement in the southwestern US can be separated into three provinces on the basis of the ε_{Nd} values of the average crustal rocks at the time of the Mesozoic and Tertiary magmatism. The relative consistency of the ε_{Nd} values over large geographic areas has been confirmed by measurements of the basement rocks themselves (Bennett and DePaolo 1987; Farmer and DePaolo 1983, 1984; Nelson and DePaolo 1985) and is a consequence of the uniformity of the Sm/Nd ratio and the crust-formation ages of the rocks from each area. The data on these granitoids and the relationship to the basement crust-formation age and ε_{Nd} values are shown in Fig. 9.7b. In each province, the rocks

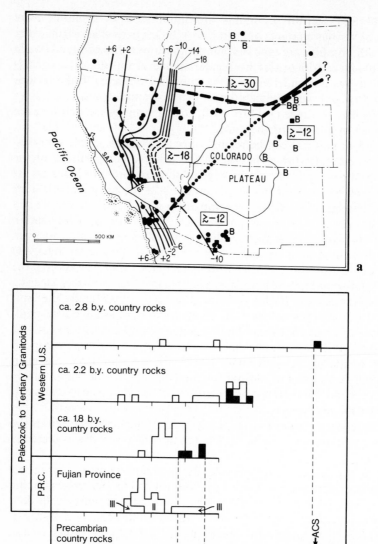

Fig. 9.7. (a) Map of the western US with ε_{Nd} contours; **(b)** ε_{Nd} histogram of granitic rocks from western US (Farmer and DePaolo 1984). Solid squares are for peraluminous granite. Data from southeast China (Xuan and DePaolo 1984) shown for comparison

emplaced near the western edge of the Precambrian basement have a large range of ε_{Nd} values that encompasses the values for primitive island-arc rocks. However, each province has a characteristic lower-limiting ε_{Nd} value for the granitoids. The granitoids of central and southern Arizona, which are emplaced into basement rocks of ca. 1.8 b.y. age, have a lower limit value of $\varepsilon_{Nd} = -10$ to -12. The granitoids of the Sierra Nevada and the northern Great Basin have a lower limit value of -17 to -19, and those of the Archean province have a lower limit value of -30. The regional synthesis unambiguously shows that the concentration of ε_{Nd} values near a lower limiting value that matches the typical local crustal values indicates that the granitoids represent magmas derived entirely from the continental crust. In this case it has even been possible to identify a previously unknown crustal province on the basis of the lower limit ε_{Nd} value (Figs. 5.12 and 5.13).

The relationship of chemical composition to isotopic composition in granitic rocks is often complicated, because they are extremely differentiated magma types. The chemical and isotopic compositions are partly decoupled during the magmatic evolution (Sect. 9.4), so it is necessary to construct relatively complex petrogenetic models in order to reconcile the chemical and isotopic data. Figure 9.8 shows the results of some model calculations (Farmer and DePaolo 1983), using the model of assimilation accompanied by fractional crystallization (DePaolo 1981 d), with assumptions having been made about the properties of the mantle-derived magma that initiates the process, and the contaminants. The relative amount of assimilation in comparison to crystallization is treated as the adjustable parameter. The calculations can reproduce the properties of the rocks that have been interpreted as forming by this process under the restrictions that (1) the rate of assimilation be less than half the rate of crystallization during the time that the assimilation was taking place, and (2) plagioclase was not an important fractionating phase during the assimilation-fractional crystallization process.

For rocks that are pure crustal melts, somewhat simpler models may be applicable (Fig. 9.9). The ε_{Nd} and the age of the probable crustal source rocks can be used to estimate the Sm/Nd ratio, and from this the shape of the rare-earth element abundance pattern of the magma source [see Eq. (3.14) and Chap. 4]. Using the parameters given in Chapter 4, Fig. 4.5b, it can be deduced that 10 to 20% partial melt of a source rock with the mineralogy of garnet granulite will have the rare-earth element and isotopic characteristics of the granitic rocks of the eastern part of the northern Great Basin.

Data from the granitic rocks of a large Paleozoic batholith in Australia have been reported by McCulloch and Chappell (1982) (Fig. 9.10). The ε_{Nd}-ε_{Sr} data are comparable to those from the southwestern US. A curved array is observed which extends from ε_{Nd} and ε_{Sr} values near the primitive arc field to values typical of Precambrian crust. The ε_{Nd} values reach a fairly well-defined lower limit value of -9 to -10. The authors interpreted the data in terms of the chemical classification of granitic rocks (I and S types) as defined by Chappel and White (1974). They found that the two types of granitic rocks

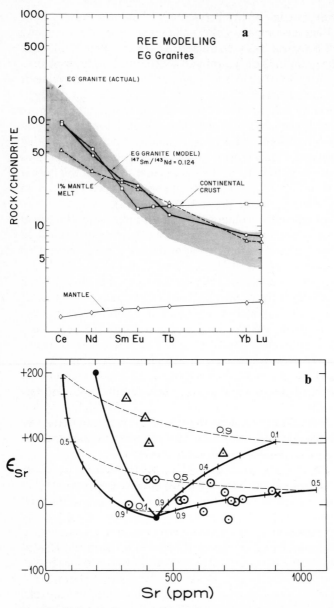

Fig. 9.8. (a) Sr isotopic composition and concentration of granitic rocks from the northern Great Basin (Farmer and DePaolo 1984; DePaolo and Farmer 1984). *Unlabelled solid line* connects the composition of a model mantle magma (*lower filled circle*) with that of a model contaminant (*upper filled circle*) and represents the trace of a "simple mixing" curve, or assimilation-fractional crystallization (AFC) with a value for $D_{Sr} = 1$. Other solid lines are calculated mixing trajectories for values of $\dot{M}_a/\dot{M}_c = 0.5$ and, clockwise from the left, $D_{Sr} = 2$, 0.5, and 0.01. Numbers on the solid lines refer to values of F, the mass of remaining magma as a fraction of the mass of the original magma. Numbers on the dashed lines refer to values of M_a/M_m, the ratio of the total

Fig. 9.9. Results of models of rare earth element concentrations in granitic rocks from the northern Great Basin that from isotopic data have been shown to be purely melted from the crust (*CG* cratonal granitoids; Fig. 9.6; from Farmer and DePaolo 1984). Magma source rare earth pattern is the same as the contaminant shown in Fig. 9.8b. The concentrations in the magma depend critically on the amount of garnet in the magma source (see Chap. 4)

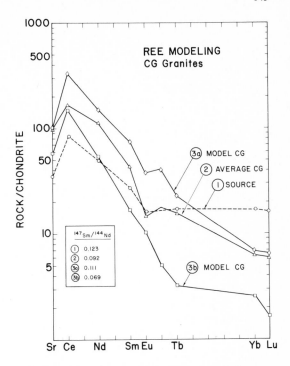

generally have different isotopic characteristics, but that there is overlap and in total they appear to be endmembers in a continuum of isotopic compositions.

The similarity of the isotopic patterns observed in the batholiths of southeastern Australia and the southwestern US suggest that the petrogenetic processes operating were similar. McCulloch and Chappell (1982), however, did not favor the assimilation-fractional crystallization model. They suggested that all of the granitic magmas were formed by intracrustal melting, and that the crustal magma sources had the different isotopic compositions observed in the granitic rocks. On the other hand, to the extent that the prebatholithic crustal structure is known, the data also are compatible with models involving mantle-derived magma. For example, the New England batholith is intruded mainly into a region of abundant Devonian or older volcanogenic sedimentary rocks, which extend inland from the location of the batholith for some dis-

◄───

mass assimilated to the original magma mass. The low ε_{Sr} values with high Sr concentrations require low values of $\dot{M}_a/\dot{M}_c$ (in some cases lower than the value of 0.5 used for this example) and a low value for D_{Sr}. The latter implies that plagioclase was not the major fractionating mineral during the AFC process. Modelling of Nd concentrations and isotopic compositions results in a similar conclusion. **(b)** Results of modelling of rare earth element concentrations in granitic rocks using the AFC model (*EG* eugeoclinal granitoids, Fig. 9.6). The initial magma is a 1% partial melt of the mantle (REE pattern shown); the contaminant is the continental crust (pattern shown). The *shaded area* shows the range of measured values

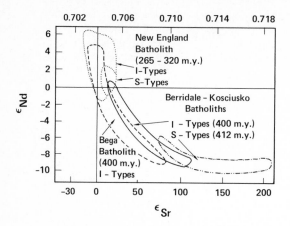

Fig. 9.10. Nd and Sr isotopic data, plutonic rocks of the Paleozoic batholiths of southeastern Australia (McCulloch and Chappell 1982)

tance. This setting is similar to that of the westernmost parts of the batholiths of the western US. The relatively restricted range of ε_{Nd} values ($+6$ to -1) could be a reflection of relatively thin crust containing no Precambrian basement, a relatively young provenance of the sedimentary rocks, and a high magma flux from the mantle. The I types could be explained as being the end products of magmas derived from the mantle that assimilated substantial amounts of the sedimentary crustal rocks during ascent. The S types may be derived from Early Paleozoic or Late Proterozoic basement or directly from melting of the sedimentary rocks.

The Bega, Berridale, and Kosciusko batholiths represent a roughly east-west transect through the Lachlan fold belt, and as such are qualitatively analogous to the transect through the northern Great Basin. The Bega batholith is closest to the edge of the continent, is composed mainly of I types, and consequently is in many ways analogous to the Sierra Nevada batholith. Most of the granitic rocks probably represent the complex mixtures formed by assimilation and fractional crystallization as mantle-derived magma passed through preexisting crust that contained both Precambrian basement and sedimentary rocks. The Berridale and Kosciusko batholiths have I-type granitoids that are similar isotopically to those of the Bega batholith, but that have assimilated more crustal rock. The S-type granitoids have a narrow range of ε_{Nd}, which includes the lower limit ε_{Nd} value for the suite. By analogy with the peraluminous granitoids of the western US, these can be interpreted as having been melted from Precambrian basement rocks that had a narrow range of ages and Sm/Nd ratios. The magmas may also have assimilated pelitic sedimentary rock, which would have enhanced the peraluminous character. The lower limit ε_{Nd} value of about -9 for the granitoids of the Berridale and Kosciusko batholiths would correspond to a lower limit ε_{Nd} value of -12 for granitic rocks of Late Cretaceous age. The data therefore suggest that the Precambrian basement in southeastern Australia in the area of these batholiths is identical in age to the ca. 1.8 b.y. basement of the southwestern US.

Fig. 9.11. (a) ε_{Nd} versus SiO_2 for Miocene volcanic rocks of the Woods Mountains Volcanic Center. **(b)** $^{87}Sr/^{86}Sr$ versus SiO_2 for Woods Mountains lavas (Musselwhite et al. 1987)

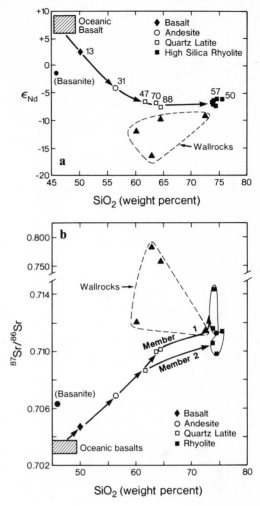

9.4 Evolution of Silicic Magma Systems

Data from Tertiary silicic volcanic centers in the western US provide insight into the generation of granitic magmas by allowing the entire sequence of lavas, ranging from basalt to high silica rhyolite, to be sampled from the same center. Examples of these are shown in Figs. 9.11 and 9.12. In both of these cases the crust is mainly Proterozoic in age with Mesozoic plutons intruding it. At the time of the Tertiary volcanism, the Nd isotopic composition of the crust and the underlying mantle was greatly different, so the interaction of mantle-derived magma with the crustal rock can be documented in some detail.

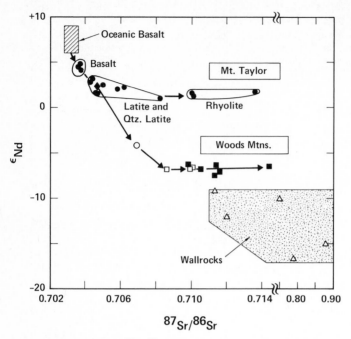

Fig. 9.12. ε_{Nd} versus $^{87}Sr/^{86}Sr$ for volcanic rocks from the Woods Mountains Volcanic Center, California, and Mt. Taylor, New Mexico (Musselwhite et al. 1987; Perry et al. 1987)

Figure 9.11 a and 9.11 b show the isotopic composition changes at Woods Mountains volcanic center in California as a function of the silica content of the magmas (Musselwhite et al. 1987). Basalt and andesite lie close to a mixing line between the probable mantle isotopic composition and the lower crustal rocks. This is interpreted as indicating that the mafic magma entering the lower crust from the mantle incorporates relatively large quantities of hot lower crustal rock, enough so that the process of admixing crustal material to the magma overshadows the differentiation by crystal fractionation processes. In the composition range andesite to high silica-rhyolite, however, there is very little shift in the ε_{Nd} value. This indicates that during the evolution of the more silica-rich magmas, assimilation of crustal rock becomes a relatively minor process in comparison with crystal fractionation. This is probably because the more siliceous magmas evolve in the shallower levels of the crust where the wallrocks are cooler and thus more heat is required for assimilation. The $^{87}Sr/^{86}Sr$ ratio, unlike the ε_{Nd} value, does change substantially in going from intermediate to silicic compositions (Fig. 9.11 b). The difference in behavior between the two systems is due to the fact that the more differentiated magmas have low Sr concentration, but high Nd concentrations, and are thus susceptible, even with only small amounts of assimilation, to shifts in $^{87}Sr/^{86}Sr$ but not to shifts in ε_{Nd}.

The data from the Woods Mountains and also those from the Mt. Taylor volcano in New Mexico are summarized on the ε_{Nd}-$^{87}Sr/^{86}Sr$ diagram (Fig. 9.12). The Mt. Taylor lavas show the same general trends as the Woods Mountains eruptives, but the total shift in ε_{Nd} is much smaller for the former. Thus overall, at Mt. Taylor the ratio of the assimilation rate to the crystallization rate was smaller than at the Woods Mountains. In both cases the final silicic magmas are clearly the result of a complex assimilation – fractional crystallization process, and are therefore crust-mantle mixtures rather than either pure differentiates of mantle magmas or pure products of crustal anatexis.

The data from the Tertiary volcanic centers help to explain some of the data on the granitic rocks. They confirm that silicic magmas form by interaction of mantle magmas with crustal rocks. Furthermore, the lack of shift in the ε_{Nd} value in the andesite-to-rhyolite sequence helps explain why the granitic rocks show such regular regional trends regardless of whether quartz diorites, granodiorites, or granites are sampled. The data also show that Sr isotope ratios are unlikely to show the same regularity because of the variability of the $^{87}Sr/^{86}Sr$ ratio in the more differentiated magma types.

Chapter 10   Continental Mafic Rocks

10.1 Flood Basalts

The interpretation of the isotopic data on continental basaltic rocks is complicated by factors that are similar to those associated with the continental magmatic arcs. In contrast to the continental margin arcs, however, intermediate composition lavas are relatively rare in comparison to basalt. Because it is more difficult in general to reconcile crustal contamination with the chemical composition of basalt, the basalt provinces present some unique problems of interpretation. These provinces are separated into three types here for the purpose of discussion. Flood basalt provinces are large volumes of lava generally having compositions that are tholeiitic or transitional to alkali basalt (Turner and Verhoogen 1960). A second type of province is the continental rift, where the lavas are usually more alkalic and of smaller volume. The third type, which is not entirely distinct from the second, comprises mafic lavas of alkalic and often extremely silica-undersaturated compositions, which are erupted in small volumes over broad areas, generally in the form of cinder cones with small numbers of associated lava flows.

A relatively large number of data are available for the flood basalt provinces of the Columbia River plateau (Carlson et al. 1981); the Scottish Hebrides, East Greenland, and Baffin Island (Carter et al. 1978b, 1979); and the Karroo province (Hawkesworth et al. 1982b). As with most other continental volcanic rocks, the ε_{Nd} and ε_{Sr} values vary from values characteristic of primitive oceanic basalts to values similar to those of Precambrian crustal rocks. However, the nature and extent of these variations differ substantially among these provinces.

A clear example of crustal contamination is provided by the Scottish Hebrides province (Fig. 10.1). The isotopic values range from values within the mantle array, to values that deviate greatly from the mantle array toward negative ε_{Nd} and negative ε_{Sr} values. The trends of the arrays are similar to those that would be expected for the mixing of mantle-derived magma with the granulite facies and amphibolite facies metamorphic rocks of the crust through which they were erupted. The existence of crustal rocks with very large negative ε_{Nd} values (Hamilton et al. 1979b) combined with the deviation of the data from the oceanic mantle array, provide a strong case for crustal contamination. The data also indicate that the granulite facies rocks were somewhat more commonly the contaminant than were the amphibolite

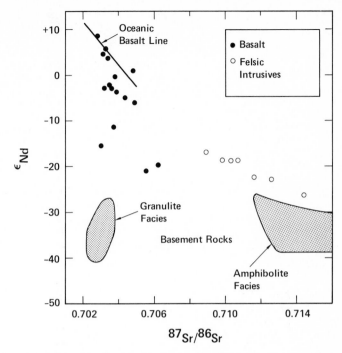

Fig. 10.1. Neodymium and strontium isotopic data for Tertiary basaltic lavas, intrusives, and Precambrian basement rocks from the Scottish Hebrides (Carter et al. 1978b)

facies rocks. This is expected because the granulite facies metamorphic rocks probably occupy the lower part of the crustal lithologic section, and therefore are likely to have been at a higher temperature and therefore more easily assimilated. Further evidence of extensive crustal melting is provided by the felsic intrusive rocks that occur in the same area with the mafic lavas and mafic intrusions, and which have isotopic compositions indicating that they were mostly melted from the crust.

The lavas of East Greenland and Baffin Island have isotopic patterns that are quite different from those of the Scottish Hebrides, even though they are a geographically related province when the effects of the Cenozoic opening of the North Atlantic ocean are restored. The Greenland lavas show very little evidence for crustal contamination insofar as all of the data lie within the mantle array (Carter et al. 1979). The lavas were erupted through Early Archean basement that probably has isotopic characteristics similar to the basement in the Scottish Hebrides, but do not appear to have been as severely affected by contamination. Carter et al. used these data to argue that the mantle sources for the lavas have isotopic compositions that are similar to those of oceanic basalts, and therefore, that the mantle beneath the continents and the oceans is similar both in its present properties and in its geologic history.

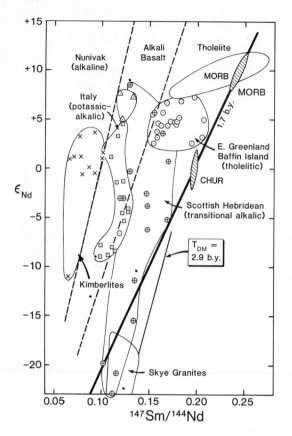

Fig. 10.2. Plot of ε_{Nd} versus $^{147}Sm/^{144}Nd$ for lavas from some continental volcanic provinces. Fields labeled *MORB* and *CHUR* are the estimated values of typical mid-ocean ridge mantle magma sources and a primitive (chondritic-Sm/Nd) magma source, respectively. Fields for *tholeiite* and *alkali basalt* are from Fig. 8.1 c. Data are from Carter et al (1978 b, 1979), Hawkesworth and Vollmer (1979), Menzies and Murthy (1979), DePaolo (1978 b), Kramers et al. (1981), Basu and Tatsumoto (1979), and Smith (1983)

A different perspective with regard to the crustal contamination problem is provided by the Sm/Nd-ε_{Nd} relationships (Fig. 10.2). The data from the Scottish Hebrides form an array that has a large slope and transgresses the fields of tholeiitic and alkali basalts as defined by the data from oceanic islands (Fig. 8.1 b). The Scottish lavas are slightly alkalic for the most part, but many plot in the tholeiitic field or even to the right of the tholeiitic field at low ε_{Nd} values. The field of the Skye granites may be representative of magma formed by melting of the crust, probably with an admixture of mantle-derived magma. The data distribution can be understood qualitatively in terms of mixing of magma derived from a typical ocean island-type mantle source with magma formed by crustal anatexis. Because the low-ε_{Nd} crustal contaminant has an Sm/Nd ratio similar to that of the high-ε_{Nd} mantle-derived magma, the mixing trend is a vertically oriented zone on this diagram. This has been noted by Carter et al. (1979) who showed that the slope of this trend corresponds to an age of 6 b.y. It is also possible to construct an alternative model for this data array. It would involve melting of low-Sm/Nd magma sources that were formed ca. 3 b.y. ago at the time of formation of the Lewisian crust. The locus of the compositions of such magma sources is shown by the model isochron on the diagram, which corresponds to a T_{DM} model age of 2.9 b.y.

Independent of origin, these transgressive data arrays on the ε_{Nd}-Sm/Nd diagram are common in continental basalts but are not observed in oceanic areas. They clearly indicate that the origin of low-ε_{Nd} magmas is different in continental and oceanic areas. The systematics of the oceanic basalts on the ε_{Nd}-Sm/Nd diagram are also valuable for assessing crustal contamination in continental basalts. For example, the data from East Greenland and Baffin Island do not show a large range of ε_{Nd}, but they are displaced to ε_{Nd} values that are significantly lower than those of the basalts from Iceland (Fig. 8.1 b). This suggests that the lavas have been contaminated with continental crust, but to a much smaller extent than those of the Scottish Hebrides.

In the Columbia River province, the basalts also exhibit a large range of isotopic compositions (Fig. 10.3). However, unlike the Hebridean lavas, the ε_{Nd}-ε_{Sr} data describe a well-defined trend and fall uniformly within the mantle array or to the lower right in a manner similar to the data from the continental margin magmatic arcs. Carlson et al. (1981) have used this characteristic to argue that essentially all of the isotopic variations are a result of crustal contamination of magmas that originally had an isotopic composition lying within the mantle array with an ε_{Nd} of about $+8$. In support of this interpretation, they presented mixing models based on the assimilation, fractional-crystallization model. Under some circumstances these can reproduce many of the isotopic and chemical compositions of the lavas. They also noted that

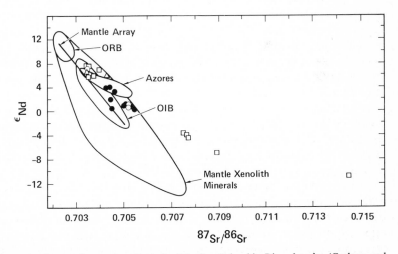

Fig. 10.3. Neodymium and strontium isotopic data for Columbia River basalts (Carlson and Lugmair 1981). *Circles* are data from the main series of lavas (Picture Gorge, Imnaha, Grande Ronde, and Wanapum basalts). *Squares* are data from the younger alkalic lavas (Saddle Mountains basalts) which are volumetrically much less abundant

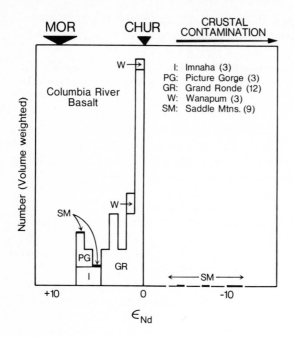

Fig. 10.4. ε_{Nd} histogram (volume weighted) for lavas from the Columbia River basalt province (DePaolo 1983a)

within the voluminous Grande Ronde series, the Mg number of the lavas correlates with ε_{Nd} and ε_{Sr} in such a way that the lavas that appear to be most contaminated with crust are also the most differentiated chemically.

An alternative interpretation of the data (DePaolo 1983a) is that most of the isotopic variations represent heterogeneity in the mantle sources. The vast majority of the lavas (99%) have ε_{Nd} values that lie within the range $+8$ to 0, and within the mantle array. This feature allows the possibility that the magma sources are similar to those of oceanic island basalt, and that crustal contamination need not have been extensive in order to explain the isotopic variations. Another interesting aspect of the data set is that, if the ε_{Nd} values are weighted according to lava volume, there appears to be a sharp cutoff at the value $\varepsilon_{Nd} = 0$, with only a tiny amount of lava having negative values (Fig. 10.4). This is consistent with a model in which the bulk of the magma came from a source having the primitive (CHUR) Sm-Nd isotopic characteristics.

The Columbia River lavas with ε_{Nd} values closer to $+8$ were mostly erupted from vents that are geographically separate from those that supplied the lavas with near-zero ε_{Nd} values. The source of the lavas with $\varepsilon_{Nd} = 0$ was apparently beneath the thicker Precambrian craton, whereas the source of the lavas with $\varepsilon_{Nd} = +8$ was beneath the thinner Paleozoic to Mesozoic crust. This arrangement also seems consistent with the assimilation hypothesis, but unlike the Hebridean province, there is a lack of the associated silicic magmatism that would be expected if crustal melting were substantial.

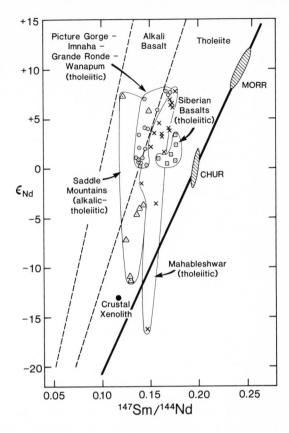

Fig. 10.5. Plot of ε_{Nd} versus $^{147}Sm/^{144}Nd$ for lavas from some flood basalt provinces. Other symbols as in Fig. 10.2. Data are from Carlson and Lugmair (1981), Mahoney et al. (1982), and DePaolo and Wasserburg (1976b, 1979b). Crustal xenolith datum is from a Saddle Mountains lava (Carlson and Lugmair 1981)

The ε_{Nd}-Sm/Nd data from Columbia River lavas (Fig. 10.5) also show little evidence for crustal contamination. The Picture Gorge – Imnaha – Grande Ronde series, which account for 99% of the erupted lava, plot close to the transition line between tholeiitic and alkali basalt as defined by the oceanic basalt data. This position, slightly to the left of the main tholeiitic field, is about what one would expect for these lavas because they are relatively highly differentiated. The data do not transgress the petrologic-isotopic boundaries defined by the oceanic basalts, and thus can be considered to be consistent with derivation of the lavas from isotopically distinct mantle sources with little contamination. On the other hand, the younger, small volume flows of the Saddle Mountain basalts do show transgressive behaviour similar to that shown by the Hebridean basalts. Transgressive behavior is also displayed by the Deccan flood basalts of the Mahableshwar Plateau (Mahoney et al. 1982), although most of the Mahableshwar data fall well within the tholeiitic field, consistent with their petrology. The Siberian basalts (DePaolo and Wasserburg 1979b) also fall within the tholeiitic field.

Yet another type of isotopic variability is displayed by the lavas of the Karroo province in southern Africa (Hawkesworth et al. 1982b). The ε_{Nd} and ε_{Sr}

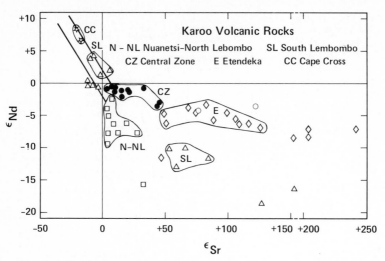

Fig. 10.6. Neodymium and strontium isotopic data for lavas from the Karroo province of southern Africa (Hawkesworth et al. 1982b)

values (Fig. 10.6) lie in a region close to the mantle array in the upper left quadrant of the diagram, but spread over a large area in the quadrant of negative ε_{Nd} and positive ε_{Sr}. The lavas from the central zone are the most voluminous and have isotopic compositions that are relatively close to $\varepsilon_{Nd} = \varepsilon_{Sr} = 0$. The petrogenesis of the lavas in this province has been interpreted by Hawkesworth et al. (1982) to have been quite complicated and variable from area to area. In one area (Nuanetsi-North Lebombo Province; N-NL) the lavas are primitive picrites that are unlikely to be contaminated with continental crust, so the isotopic variations are interpreted to be representative of mantle reservoirs beneath the continent. In South Lebombo (SL) the isotopic data and the trace element patterns show some correlation, so the isotopic variations are attributed to crustal contamination. In Cape Cross (CC) and Etendeka (E) the lavas exhibit a bimodal distribution of isotopic values. Some of the lavas appear to be derived from mantle reservoirs with characteristics similar to the sources of oceanic basalts, whereas others come from mantle that has a composition lying in the negative ε_{Nd}, positive ε_{Sr} field.

The Karroo province is unique in that it is also a locality of extensive kimberlite volcanism. This provides an opportunity to sample the underlying crust and mantle directly through the xenoliths in the kimberlites. Hawkesworth et al. found evidence for enriched reservoirs in the mantle in the form of metasomatized, K-rich ultramafic nodules that contain the mineral richterite. Both the peridotite nodules that contain richterite and those that do not have similar ε_{Nd} values between $+2$ and -4, but the richterite-bearing nodules have high ε_{Sr} values in the range of $+3$ to $+85$. These two types of nodules bracket the isotopic composition seen in the central zone basalts, but the isotopic values in the other basalts are more extreme than those of the nodules. The crustal nodules have very large negative ε_{Nd} values and positive ε_{Sr} values.

The Karroo lavas present evidence for isotopic heterogeneity in the sub-continental mantle as well as crustal contamination effects. However, the large volume of lava of the central zone has ε_{Nd} values that are very close to zero, a characteristic that is similar to that observed for the Columbia River province. It appears possible that a mantle source having primitive character-istics with respect to both Sm-Nd and Rb-Sr isotopes was important in both the Karroo and Columbia River provinces. In both cases there appears to have been mixing between this source and other mantle materials as well as crustal rock.

10.2 Extensional Provinces

The lavas of the Oslo Rift exhibit considerable complexity in their isotopic ratios (Fig. 10.7). The ε_{Nd} values fall within the range of oceanic values, but there is no correlation with ε_{Sr} values. Jacobsen and Wasserburg (1978b) in-terpret these data in terms of three source materials for the magmas. One has ε_{Nd}-ε_{Sr} values that lie below the mantle array (basalts of Skien – S) and another has ε_{Nd} values close to zero and positive ε_{Sr} values. The former is represented by silica-undersaturated nephelinite and ankaramite, whereas the latter is represented by tholeiitic basalt. The isotopic compositions can proba-

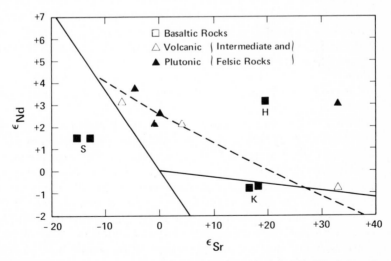

Fig. 10.7. Neodymium and strontium isotopic data for lavas from the Oslo Rift (Jacobsen and Wasserburg 1978)

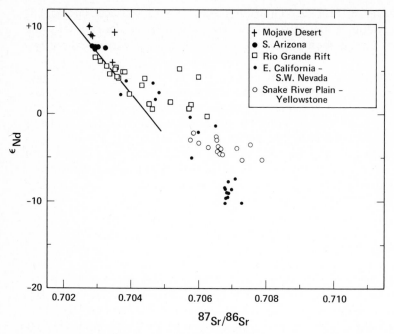

Fig. 10.8. Neodymium and strontium isotopic data for basalts from the western US (data from Menzies et al. 1983; Perry et al. 1987; Semken 1984)

bly be most easily explained in terms of crustal contamination as with the Hebridean lavas, in view of the Proterozoic continental crust in the rift area. The intermediate and silicic rocks of the rift show somewhat more systematic variations. The data form a trend from positive ε_{Nd} values near the mantle array to lower ε_{Nd} values that plot to the right of the array. This trend is also likely to be due to crustal contamination as the isotopic values correlate in the normal way with chemical composition. The primary mantle magma source for the Oslo Rift lavas appears to have had an ε_{Nd} value in the range $+4$ to $+8$.

The data from the extensive Cenozoic basalt province of the southwestern US (Menzies et al. 1983; Semken and DePaolo 1983) have many of the characteristics of the continental flood basalts (Fig. 10.8). This province is of particular interest because the magmas have perforated continental crust of varying ages and thicknesses, and are related to a variety of tectonic environments. In some areas of the southwestern US there is clear evidence for a mantle source with oceanic affinities. In addition, there are a large number of lava flows with isotopic compositions that fall within the field of Precambrian crust. In many cases the isotopic compositions cannot be explained in terms of crustal assimilation because the lavas are magnesium-rich and relatively poor in alkalies.

Menzies et al. (1983) have attributed the isotopic variations in these basalts to the heterogeneous mantle. They further hypothesize that the heterogeneity is in the subcontinental lithospheric mantle and that it is a result of the lithosphere having been isolated from the rest of the mantle since the Precambrian with crustlike values of Sm/Nd and Rb/Sr. Semken and DePaolo (1983) and Semken (1984), however, argue that because this part of the North American continent has been involved in several tectonothermal events in the Phanerozoic it is unlikely that such subcontinental mantle has survived since the Precambrian. They proposed instead that the heterogeneity was introduced more recently, in conjunction with the Cenozoic magmatism or as a result of the Mesozoic subduction at the continental margin.

The idea of low-ε_{Nd} subcontinental mantle was taken a step further by Perry et al. (1987). In a study of isotopic variations in basalts from the Rio Grande rift, they found that the ε_{Nd} values of alkali basalts were systematically related to position relative to the rift axis, and also decreased from south to north in the rift. They interpreted the results in terms of upwelling high-ε_{Nd} mantle in the rift region and its replacing of the preexisting lower-ε_{Nd} lithospheric mantle. Tholeiites and alkali basalts were used as samplers of the mantle at different depths; the tholeiites originating at smaller depth than the alkali basalts.

One of the most important results of the studies on continental alkali basalts is that the isotopic composition of the most common subcontinental mantle magma sources is similar to those of typical oceanic alkali basalts from intraplate islands (also see Zindler and Hart 1986). In continental areas there is the added complication that some of the lavas are contaminated with continental crust or subcontinental mantle, but where estimation of the characteristics of uncontaminated lava can be made, the ε_{Nd} value is generally in the range $+6$ to $+8$ and the ε_{Sr} value is in the range -10 to -20. So there is ample evidence that the sources of alkali basalts are similar in both continental and oceanic areas. The areas of volcanism in both cases are also underlain by seismic low-velocity zones at depths of 80 to 150 km, and the chemical composition of the lavas is relatively consistent. Models of mantle structure must incorporate the fact that the thermal, rheological, chemical, and isotopic properties of the mantle supplying alkali basalts are similar worldwide.

10.3 Kimberlites and Ultrapotassic Lavas

Kimberlites provide a potentially valuable probe of mantle isotopic compositions because they are believed to originate deep in the mantle (more than 150 km). The Nd and Sr isotopic data for kimberlites and related rocks are shown in Fig. 10.9. The data reported by Basu and Tatsumoto (1979) all have ε_{Nd} values within about 2 units of the chondritic values. They interpreted their data as indicating that the magmas were derived from primitive mantle source regions with a chondritic rare-earth element composition. However,

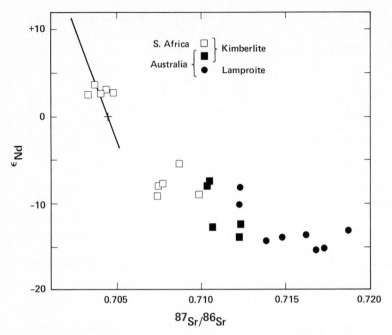

Fig. 10.9. Neodymium and strontium isotopic data for kimberlites and other related rocks (data from DePaolo 1978b; Basu and Tatsumoto 1979; Kramers et al. 1981; McCulloch et al. 1983; Smith 1983)

data reported by DePaolo (1978b), Kramers et al. (1981), McCulloch et al. (1983), Smith (1983), and Basu et al. (1984) are not consistent with the hypothesis of Basu and Tatsumoto. The other data, from Cretaceous kimberlites of South Africa and Australia have highly variable ε_{Nd} and ε_{Sr} values. These data clearly show that kimberlites are not derived from primitive mantle reservoirs. The kimberlite source regions in the mantle are heterogeneous, but apparently of two general types (Smith 1983). One source (with positive ε_{Nd}) is only slightly different from the source of typical oceanic alkali basalts. The other source, represented by the large number of samples with negative ε_{Nd} values, must be regions of the mantle that are particularly rich in old recycled crustal materials. In fact, all of the data could be explained as representing regions of the mantle containing different proportions of recycled crust. This explanation is particularly attractive because of the coherent trend observed on the ε_{Nd}-ε_{Sr} diagram, and the fact that the lower limit ε_{Nd} observed corresponds to an age of about 2 b.y., which is close to the average age of the crust. The only alternative is that the sources are mantle materials that acquired low Sm/Nd and high Rb/Sr ratios about 2 b.y. ago and have survived destruction by mantle convection since then. Considering that the apparent mixing time of the mantle is of the order of 0.5 b.y. as discussed earlier, this

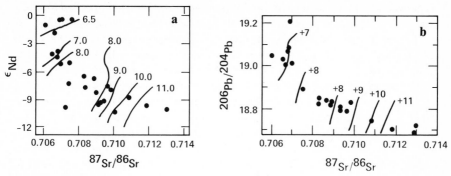

Fig. 10.10. (a) Nd and Sr isotopic ratios of potassic volcanic rocks from the Roman region showing location of $\delta^{18}O$ contours. **(b)** Pb, Sr, and O isotopic data from lavas of the Roman region (Taylor et al. 1979)

seems unlikely unless the source regions were in the continental lithosphere for most of this time.

An extensive study of potassium-rich volcanic rocks from Italy has been reported by Hawkesworth and Vollmer (1979) (data included in Fig. 10.10). This province is a classic one for the study of the petrology of ultrapotassic lavas. The rocks exhibit a large range of ε_{Nd} and ε_{Sr} values and strong geographic regularities. From north to south in the province there is a regular decrease in the ε_{Sr} values and an increase in the ε_{Nd} values. The rocks with the highest ε_{Sr} values are silica-rich, whereas most of the other rocks are low in silica and anomalously high in potassium. As shown by Taylor et al. (1979), the strontium and neodymium isotopes correlate with oxygen and lead isotope ratios and with the chemical composition of the lavas (Fig. 10.10a, b). Taylor et al. (1979) have argued that a substantial amount of the variations can be attributed to assimilation of sedimentary wallrocks that had high $\delta^{18}O$, high ε_{Sr}, low ε_{Nd}, and low $^{206}Pb/^{204}Pb$. To explain the chemical data it is required that fractional crystallization accompanied the assimilation and that plagioclase was the primary fractionating mineral phase. The assimilation and fractional crystallization is postulated to have occurred at relatively shallow crustal levels. Additional crustal interaction at deeper levels also appears possible. The authors suggest that the apparently higher proportion of assimilated crust in the northern part of the province (Tuscany) was due to prior heating of the crust by an earlier magmatic event. Hawkesworth and Vollmer (1979) prefer to interpret the lavas from the southern parts of the Italian province as partial melting products of mantle reservoirs that already had low ε_{Nd} values and high ε_{Sr} values. They postulate that these mantle regions acquired these characteristics during a metasomatic enrichment event that predated the magmatism by several hundred million years.

Despite the abundant evidence for crustal interaction and the geographic variations of the isotopic ratios, it is still not clear why the lavas are so rich in potassium. The data clearly show that K-enrichment accompanied assimilation and fractional crystallization, but that the latter was the determining factor for most of the chemical characteristics. It seems necessary that the parental magmas were already rich in potassium, a conclusion that lends credence to the enriched mantle model of Hawkesworth and Vollmer. However, Taylor et al. also point out that all of the parental high-K magmas had $\delta^{18}O$ values that were significantly higher than typical mantle values. It is not possible to cause ^{18}O enrichment by mantle metasomatism if the metasomatism involves only transfer of fluid from one "normal" region of the mantle to another. The high $\delta^{18}O$ values could be attributed to deeper crustal interaction. In any case, the low ε_{Nd} values clearly indicate a source material with low Sm/Nd, and thus probably enriched in light rare-earth elements. A LREE-enriched source material would help to explain the very high concentrations of these elements in the lavas.

The ε_{Nd}-Sm/Nd data on the Italian lavas and kimberlites also are shown in Fig. 10.2. The Italian lavas, which are highly undersaturated with respect to silica, show slight transgressive behavior as is characteristic of other continental provinces that exhibit large variations in ε_{Nd}. Kimberlites, on the other hand, fall well within the field of silica-undersaturated lavas, which is consistent with their being derived from mantle sources with characteristics that are similar to, but more extreme than, oceanic alkali basalts. The kimberlites appear to be highly fractionated relative to their mantle sources with respect to their Sm/Nd ratios. The kimberlite and lamproite data (Fig. 10.9) fall on an extension of the data array shown in Fig. 8.1. This is consistent with the correlation noted between silica undersaturation, low $a_{Sm/Nd}$ values, and low ε_{Nd} values, and supports the hypothesis that admixed recycled crust in the mantle is responsible for the low ε_{Nd} values of the kimberlites. The Italian province provides an additional example, albeit not as clear as the Scottish Hebrides, where crustal contamination is associated with transgressive behaviour in the ε_{Nd}-Sm/Nd characteristics.

10.4 Layered Mafic Intrusions

Layered mafic intrusions are of particular interest because they provide a means of studying the processes in mafic magma chambers. The cumulate rocks provide a record of the crystallization history of the magma. Isotopic ratios in these rocks give information on the interaction of the magma body with its wallrocks, such as the rate at which assimilation of wallrock progressed in comparison to the rate of crystallization and cooling of the magma, and the degree to which inhomogeneities introduced by assimilation were homogenized by convection. This type of information is fundamental for understanding all isotopic data on igneous rocks and their relationship to the properties of the magma sources.

Fig. 10.11a,b. Neodymium and strontium isotopic data for the Kiglapait Intrusion, Labrador. **(a)** ε_{Nd} versus An content of plagioclase; **(b)** $^{87}Sr/^{86}Sr$ versus An content of plagioclase

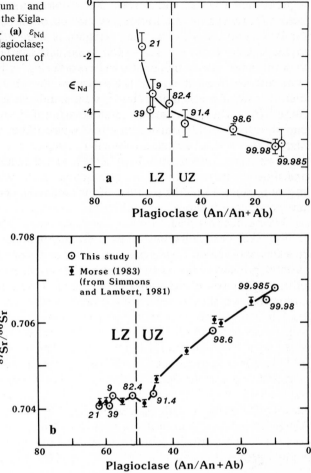

One example has been given previously (Fig. 2.3). The Sm-Nd isotopic data on the Stillwater intrusion indicate that the magma was homogeneous with respect to ε_{Nd} over a stratigraphic interval of several kilometers, representative of about two-thirds of the crystallization products of the magma. A different result was obtained for the Kiglapait intrusion of Labrador (Morse 1969). The Kiglapait intrusion differs from the Stillwater in several ways, one of which is that the top and bottom of the Kiglapait are exposed. The upper part of the Stillwater intrusion is covered by younger sedimentary rocks. Through a combination of mapping and gravity studies, the shape of the Kiglapait intrusion can be modeled so that the cumulate stratigraphy can be correlated with the percent of magma crystallized (Morse 1969). The data shown in Fig. 10.11 is keyed to the stratigraphic volume-solidified index (denoted PCS for percent solidified).

Both the initial ε_{Nd} and the initial $^{87}Sr/^{86}Sr$ values vary in a more or less regular fashion from the bottom to the top of the cumulate series in the Kiglapait. For $^{87}Sr/^{86}Sr$ the total variation is from 0.70408 to 0.70679; for ε_{Nd} the range is from -1.6 to -5.6. These data can be understood in terms of progressive contamination of the crystallizing magma (Chap. 4). With some assumptions about the properties of the assimilated wallrock, the relative rates of assimilation and crystallization can be estimated (DePaolo 1985). The Kiglapait magma assimilated small amounts of wallrock ($\dot{M}_a/\dot{M}_c = 0.02$ to 0.07) throughout its crystallization history. The irregular progression of $^{87}Sr/^{86}Sr$ values in the lower parts of the intrusion suggest that there were injections of fresh (low $^{87}Sr/^{86}Sr$) magma added to the chamber as crystallization proceeded. The smooth isotopic shifts in the upper parts of the intrusion indicate that the addition of uncontaminated magma was unimportant in the late stages of crystallization, but that assimilation continued as crystallization proceeded.

The Skaergard intrusion of East Greenland (Wager and Brown 1967; Stewart and DePaolo 1987) provides another opportunity to study magma chamber processes. The Skaergard intrusion has a much smaller volume than either the Kiglapait or the Stillwater intrusions. The wallrock of the Skaergard intrusion is largely Archean gneiss that has very low ε_{Nd} values (typically

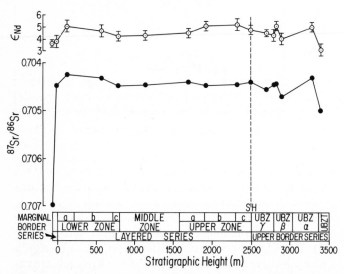

Fig. 10.12. Neodymium and strontium isotopic data for the Skaergard Intrusion, East Greenland. The $^{87}Sr/^{86}Sr$ scale is reversed so that isotopic shifts due to assimilation are in the same direction (downward) for both Nd and Sr. The data are plotted against stratigraphic height in the intrusion. The magma body crystallized from both the top and the bottom. The Sandwich Horizon (*SH*) represents the last magma to crystallize. The isotopic variations are small, but are seen for both Nd and Sr, and in both the Layered Series and the Upper Border Series (Stewart and DePaolo 1987)

− 40) that contrast strongly with those of the intrusion (about + 5). Thus, the Nd (and also Sr) isotopic composition of the Skaergard magma was senstive to small amounts of assimilation of the surrounding rock. However, the isotopic variations exhibited by the Skaergard cumulate rocks, from the lowest available cumulate rocks to the final liquid (the Sandwich Horizon), are very small (Fig. 10.12). Nevertheless, these small variations correlate well with more easily resolved variations in the Sr isotope ratio, and appear to be indicative of both assimilation and some magmatic recharge in the system. The Upper Border Zone of the intrusion, and particularly the Marginal Border Series, show the largest isotopic shift from assimilation. The Lower and Middle Zone rocks show the effects of progressive contamination, and this trend is mirrored in the Upper Border Series rocks. This indicates that there was whole-chamber convection, or at least good chemical communication throughout the chamber during the first half of the crystallization of the intrusion. However, the last crystallized liquid in the Sandwich Horizon appears to be less contaminated than the Middle Zone, suggesting that a pulse of uncontaminated magma entered the chamber during the accumulation of the Middle Zone cumulates. The overall low assimilation rates of the Skaergard system could be related to the small size of the magma chamber and the shallow depth of intrusion. The cold wallrocks apparently caused a chilled marginal zone to form that acted to largely isolate the magma from interaction with them.

10.5 Ultramafic Nodules

Nodules of mantle peridotite contained within alkalic lavas and kimberlites provide an alternative means of examining isotopic variations in the mantle. This method complements the data from lavas because it is a finer-scale sampling method. Strictly speaking, it is necessary to know the isotopic variability as a function of scale in order to assess the mechanical homogenization of the mantle by convection. On the other hand, it is not clear that lavas and nodules sample the same parts of the mantle.

The isotopic data on most of the ultramafic nodules that have been studied have been measured on separated minerals rather than on whole rock specimens (Fig. 10.13). Analyses of whole rocks have not been made because of the findings of several investigators showing that fine-grained materials associated with grain boundaries and fractures contain a substantial fraction of the incompatible elements (e.g. Zindler and Jagoutz 1987). It is possible that these fine-grained materials were introduced during the transport to the surface while the nodule was entrained in a high-temperature magma. Consequently, most investigators have chosen to avoid the potential complications by analyzing separated minerals that are relatively rich in incompatible elements. In general, almost the entire budget of Sr and Nd is contained in the mineral clinopyroxene (or amphibole, if present). These minerals concen-

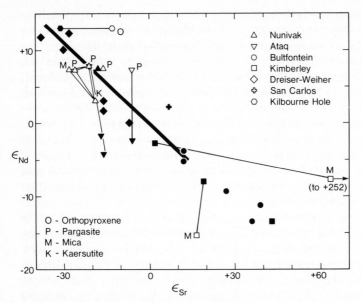

Fig. 10.13. Neodymium and strontium isotopic data for ultramafic nodules from kimberlites and alkali basalts (data from Menzies and Murthy 1979, 1980a; Stosch et al. 1980)

trate Sr and Nd by a factor of 50 or more relative to olivine, orthopyroxene, and spinel.

The ε_{Nd} and ε_{Sr} values of mantle pyroxenes do not vary greatly from the mantle array and its extension to larger negative ε_{Nd} values. However, in nodules where more than one mineral has been analyzed, it is common to find large differences in ε_{Nd} or ε_{Sr} between minerals. One example is provided by the nodules from the Ataq alkali basalts in South Yemen, which contain diopside and pargasitic amphibole. The amphibole has an ε_{Nd} value of about $+7$, whereas the diopside has ε_{Nd} of -2 to -4. In another nodule, the pargasite has ε_{Nd} of $+7$, similar to mica from the same nodule, and kaersutitic amphibole has ε_{Nd} of $+3$. Other examples of mineral-scale heterogeneity are seen in clinopyroxene-mica pairs from the Kimberly pipe and a clinopyroxene-orthopyroxene pair from Kilbourne Hole, New Mexico.

In some cases the isotopic composition of the nodule minerals are related to the isotopic composition of the host lavas, but in other cases they are completely different. At Ataq and at Nunivak Island the mantle minerals have isotopic compositions that are similar to the lavas. For the Ataq suite, bimodality in the nodule data corresponds to bimodality in the data from the lavas. On the other hand, the lavas of Dreiser-Weiher are consistently different in isotopic composition, in some cases by as much as 12 units of ε_{Nd}, from the included nodules. Similarly, The ε_{Nd} values of kimberlites from Bultfontein, South Africa, are much different from the included nodules. The nodules at Kimberley and Bultfontein are interesting in that the diopsides have well-correlated variations of ε_{Nd} and ε_{Sr}, which describe an almost perfect extension to the mantle array defined by data from oceanic lavas.

The significance of the nodule data vis-à-vis the data on lavas is not clear. A particular problem is that the nodules must of necessity come from a level of the mantle that is shallower than the source of the lavas. The long-lived, fine-scale isotopic heterogeneity preserved in the nodules is testimony to the fact that they were stored in the mantle at temperatures substantially below those necessary to produce partial melting. In most cases they come from continental regions where the surface rocks are of Precambrian age. Thus, most of them are from the subcontinental lithosphere, and therefore represent rock material that has been separated from the main volume of convecting upper mantle material for some time.

The subcontinental mantle, although sharing some of the genetic features of the continental crust, differs from the continental crust isotopically because it does not have a narrow range of Sm/Nd ratios. The Sm/Nd ratios vary from much less than to much greater than the chondritic value, with ε_{Nd} values to match. The isotopic structure deduced for typical Precambrian lithosphere is one of (1) crust with ε_{Nd} values that are large negative numbers, proportional to age, and (2) underlying mantle lithosphere with ε_{Nd} values that vary from positive values to a lower limit close to the average overlying crustal values. This model was previously proposed by Basu and Tatsumoto (1979) and is inherent in the general model depicted in Fig. 7.11.

The existing data do not allow any general conclusion to be reached regarding the precise chemical composition, and especially the radiogenic heat production of typical subcontinental lithosphere. Many nodules from continental areas show enrichment in light rare-earth elements, low Sm/Nd ratios, low ε_{Nd} values, and correspondingly high concentrations of uranium, thorium, and potassium (Hawkesworth et al. 1984; Frey 1982). However, in the same areas, other nodules have just the opposite features. It is not possible to determine the relative amounts of the two types of ultramafic rock in the subcontinental lithosphere. Nevertheless, it is evident that the subcontinental lithosphere is not composed entirely of very depleted harzburgite as was once thought.

The overall variability of the ε_{Nd} and ε_{Sr} values of the nodules is also worthy of comment. Somewhat surprisingly, most of the nodule data lie close to the mantle array. As noted before, it is not clear that the nodules can be taken to be representative of the oceanic mantle similar to the sources of the oceanic basalts. However, if they were, it would imply that the fractionation processes that have affected the mantle have always produced Sm/Nd and Rb/Sr variations such that the evolving ε_{Nd} and ε_{Sr} values do not depart significantly from the mantle array. Therefore, the narrowness of the mantle array might be less surprising than suggested in Chapter 5. However, the processes that are responsible for this coherence in the ε_{Nd} and ε_{Sr} values are still poorly defined.

References

Aldrich LT, Nier AO (1948) Argon 40 in potassium minerals. Phys Rev 74:876 – 877

Allegre CJ (1982) Chemical geodynamics. Tectonophysics 81:109 – 132

Allegre CJ, Ben Othman D (1980) Nd-Sr isotopic relationship in granitoid rocks and continental crust development: a chemical approach to orogenesis. Nature 286:335 – 342

Allegre CJ, Minster JF (1978) Quantitative models of trace element behavior in magmatic processes. Earth Planet Sci Lett 38:1 – 25

Allegre CJ, Rousseau D (1984) The growth of the continent through time studied by Nd isotope analyses of shales. Earth Planet Sci Lett 67:19 – 34

Allegre CJ, Ben Othman D, Polve M, Richard P (1979) The Nd-Sr isotopic correlation in mantle materials and geodynamic consequences. Phys Earth Planet Inter 19:293 – 306

Allegre CJ, Brevart O, Dupre B, Minster J-F (1980) Isotopic and chemical effects produced in a continuously differentiating convecting earth mantle. Philos Trans R Soc Lond A Math Phys Sci 297:447 – 477

Allegre CJ, Dupre B, Lambert B, Richard P (1981) The subcontinental versus suboceanic debate. I. Lead-neodymium-strontium isotopes in primary alkali basalts from a shield area: the Ahaggar volcanic suite. Earth Planet Sci Lett 52:85 – 92

Allegre CJ, Hart SR, Minster J-F (1983) Chemical structure and evolution of the mantle and continents determined by inversion of Nd and Sr isotopic data. II. Numerical experiments and discussion. Earth Planet Sci Lett 66:191 – 213

Anders E (1977) Chemical compositions of the moon, earth, and eucrite parent body. Philos Trans R Soc Lond A Math Phys Sci 285:23 – 40

Anderson CA, Blacet PM, Silver LT, Stern TW (1971) Revision of Precambrian stratigraphy of the Prescott-Jerome area, Yavapai County, Arizona. US Geol Surv Bull 1324-C:1 – 16

Anderson DL (1979) Chemical stratification of the mantle. J Geophys Res 84:6297

Anderson DL (1981) Hotspots, basalts and the evolution of the mantle: a model. Science 213:82

Anderson DL (1982) Isotopic evolution of the mantle: a model. Earth Planet Sci Lett 57:13 – 24

Archibald DA, Farrar E (1976) K-Ar ages of amphiboles from the Bay of Islands ophiolite and the Little Port Complex, western Newfoundland and their geological implications. Can J Earth Sci 13:520 – 529

Armstrong RL (1981) Radiogenic isotopes: the case for crustal recycling on a near steady-state no-continental-growth earth. In: Moorbath S, Windley BF (eds) The origin and evolution of the Earth's continental crust. The Royal Society, London, pp 259 – 287

Arndt NJ, Nisbet EG (eds) (1982) Komatiites. Allen & Unwin, London, 526 p

Arth JG (1976) Behavior of trace elements during magmatic processes – a summary of theoretical models and their applications. J Res US Geol Surv 4:41 – 47

Ashwal LD, Wooden JL (1983) Sr and Nd isotope geochronology, geologic history, and origin of the Adirondack anorthosite. Geochim Cosmochim Acta 45:1875 – 1885

Baadsgaard H (1976) Further U-Pb dates on zircons from the early Precambrian rocks of the Godthaabsfjord area, West Greenland. Earth Planet Sci Lett 33:261 – 267

Basu AR, Tatsumoto M (1979) Sm-Nd systematics in kimberlites and in the minerals of garnet lherzolite inclusions. Science 205:388 – 390

Basu AR, Rubury E, Mehnert H, Tatsumoto M (1984) Sm-Nd, K-Ar and petrologic study of some kimberlites from eastern United States and their implication for mantle evolution. Contrib Mineral Petrol 86:35 – 44

Bateman P, Clarke LD, Huber NK, Moore JG, Rinehart CD (1963) The Sierra Nevada Batholith. US Geol Surv Prof Pap 414-D:46

Ben Othman D, Polve M, Allegre CJ (1984) Nd-Sr isotopic composition of granulites and constraints on the evolution of the lower continental crust. Nature 307:510–515

Benjamin T, Heuser WR, Burnett DS, Seitz MG (1980) Actinide crystal-liquid partitioning for clinopyroxene and $Ca_3(PO_4)_2$. Geochim Cosmochim Acta 44:1251–1264

Bennett VC, DePaolo DJ (1987) Proterozoic crustal history of the western United States as determined by neodymium isotopic mapping. Bull Geol Soc Am 99:674–685

Boettcher AL, O'Neil JR (1980) Stable isotope, chemical, and petrographic studies of high-pressure amphiboles and micas: evidence for metasomatism in the mantle source regions of alkali basalts and kimberlites. Am J Sci 280-A:594–621

Bowen NL (1914) The ternary system diopside-forsterite-silica. Am J Sci 33:551–573

Bowen NL (1928) The evolution of the igneous rocks. Dover, New York, 334 p

Boynton WV (1975) Fractionation in the solar nebula: condensation of yttrium and the rare earth elements. Geochim Cosmochim Acta 39:569–584

Broecker WS (1963) Radioisotopes and large-scale ocean mixing. In: Hill MN (ed) The Sea, vol 2, Chapter 4. Interscience, New York

Brown L, Klein J, Middleton R, Sacks IS, Tera F (1982) ^{10}Be in island-arc volcanoes and implications for subduction. Nature 299:718–720

Cameron AE, Smith DH, Walker RL (1969) Mass spectrometry of nanogram-size samples of lead. Anal Chem 41:525–526

Carlson RW (1982) Diez de Medina, D. Sm-Nd geochronology of Archean gneisses from the southwestern Bighorn Mountains, Wyoming. Carnegie Institution of Washington Year Book 81:536–543

Carlson RW (1987) Geochemical evolution of the crust and mantle. Rev Geophys Space Phys 25:1011–1020

Carlson RW, Lugmair GW (1979) Sm-Nd constraints on lunar differentiation and the evolution of KREEP. Earth Planet Sci Lett 45:123–132

Carlson RW, Lugmair GW (1981) Time and duration of lunar highlands crust formation. Earth Planet Sci Lett 52:227–238

Carlson RW, MacDougall JD, Lugmair GW (1978) Differential Sm-Nd evolution in oceanic basalts. Geophys Res Lett 5:229–232

Carlson RW, Lugmair GW, MacDougall JD (1981) Columbia River volcanism: the question of mantle heterogeneity or crustal contamination. Geochim Cosmochim Acta 45:2483–2500

Carmichael, ISE, Turner FJ, Verhoogen J (1974) Igneous petrology. McGraw-Hill, New York, 739 p

Carter SR, Evensen NM, Hamilton PJ, O'Nions RK (1978a) Continental volcanics derived from enriched and depleted source regions; Nd and Sr isotope evidence. Earth Planet Sci Lett 37:401–408

Carter SR, Evensen NM, Hamilton PJ, O'Nions RK (1978b) Neodymium and strontium isotope evidence for crustal contamination of continental volcanics. Science 202:743–747

Carter SR, Evensen NM, Hamilton PJ, O'Nions RK (1979) Basalt magma sources during the opening of the North Atlantic. Nature 281:28–30

Chappell BW, White AJR (1974) Two contrasting granite types. Pac Geol 8:173–174

Chauvel C, Dupre B, Jenner GA (1985) The Sm-Nd age of Kambalda volcanics is 500 Ma too old! Earth Planet Sci Lett 74:315–324

Chen C-Y, Frey FA (1983) Origin of Hawaiian tholeiite and alkalic basalt. Nature 302:785–789

Clarke WB, Beg MA, Craig H (1969) Excess ^{3}He in the sea: evidence for terrestrial primordial helium. Earth Planet Sci Lett 6:213–220

Clauer N, O'Neil JR, Bonnot-Courtois C (1982) The effect of natural weathering on the chemical and isotopic composition of biotites. Geochim Cosmochim Acta 46:1755–1762

Clayton DD (1968) Principles of stellar evolution and nucleosynthesis. McGraw-Hill, New York, 612 p

Coates RR (1962) Magma type and crustal structure in the Aleutian arc. in Crust of the Pacific Basin. Am Geophys Union Monogr 6:92–109

Cohen RS, O'Nions RK (1982) The lead, neodymium and strontium isotopic structure of ocean ridge basalts. J Petrol 23:299 – 324

Cohen RS, Evensen NM, Hamilton PJ, O'Nions RK (1980) U-Pb, Sm-Nd and Rb-Sr systematics of mid-ocean ridge basalt glasses. Nature 283:149

Coleman RG (1977) Ophiolites. Springer, Berlin Heidelberg New York, 133 pp

Collerson KD, Fryer BJ (1978) The role of fluids in the formation and subsequent development of early continental crust. Contrib Mineral Petrol 67:151 – 167

Collerson KD, McCulloch MT (1982) The origin and evolution of Archean crust as inferred from Nd, Sr, and Pb isotopic studies in Labrador. Abstracts of the Fifth International Conference on Geochronology, Cosmochronology and Isotope Geology, Nikko, Japan, 61 – 62

Condie KC (1980) The Tijeras Greenstone: evidence for depleted upper mantle beneath New Mexico during the Proterozoic. J Geol 88:603 – 609

Craig H, Lupton JE (1976) Primordial neon, helium, and hydrogen in oceanic basalts. Earth Planet Sci Lett 31:369 – 385

Craig H, Lupton JE, Welhan JA, Poreda R (1978) Helium isotope ratios in Yellowstone and Lassen Park volcanic gases. Geophys Res Lett 5:897 – 900

Crouch EAC (1963) Thermal ionization of elements of high ionization potential. In: Elliot RM (ed) Advances in mass spectrometry, vol 20, pp 157 – 162

Dallmeyer RD, Williams H (1975) $^{40}Ar/^{39}Ar$ ages from the Bay of Islands metamorphic aureole: their bearing on the timing of Ordovician ophiolite obduction. Can J Earth Sci 12:1685 – 1690

Dasch EJ (1969) Strontium isotopes in weathering profiles, deep sea sediments, and sedimentary rocks. Geochim Cosmochim Acta 33:1521 – 1552

Davies GF (1981) Earth's neodymium budget and structure and evolution of the mantle. Nature 290:208

Dempster AJ (1918) A new method of positive ray analysis. Phys Rev 11:316 – 325

DePaolo DJ (1978a) Study of magma sources, mantle structure and the differentiation of the earth from variations of $^{143}Nd/^{144}Nd$ in igneous rocks. Ph D Thesis, California Institute of Technology, 360 p

DePaolo DJ (1978b) Nd and Sr isotope systematics of young continental igneous rocks. US Geol Surv Open-File Rep 78-701:91 – 93

DePaolo DJ (1979) Implications of correlated Nd and Sr isotopic variations for the chemical evolution of the crust and mantle. Earth Planet Sci Lett 43:201 – 211

DePaolo DJ (1980) Crustal growth and mantle evolution: inferences from models of element transport and Nd and Sr isotopes. Geochim Cosmochim Acta 44:1185 – 1196

DePaolo DJ (1981a) Nd isotopic studies: some new perspectives on earth structure and evolution. EOS 62:137 – 140

DePaolo DJ (1981b) A neodymium and strontium isotopic study of the Mesozoic calc-alkaline granitic batholiths of the Sierra Nevada and Peninsular Ranges, California. J Geophys Res 86:10470 – 10488

DePaolo DJ (1981c) Neodymium isotopes in the Colorado Front Range and crust-mantle evolution in the Proterozoic. Nature 291:193 – 196

DePaolo DJ (1981d) Trace element and isotopic effects of combined wallrock assimilation and fractional crystallization. Earth Planet Sci Lett 53:189 – 202

DePaolo DJ (1983a) Comment on 'Columbia River volcanism: the question of mantle heterogeneity or crustal contamination' by RW Carlson, GW Lugmair and JD MacDougall. Geochim Cosmochim Acta 47:841 – 844

DePaolo DJ (1983b) The mean life of continents: estimates of continental recycling rates from Nd and Hf isotopic data and implications for mantle structure. Geophys Res Lett 10:705 – 708

DePaolo DJ (1983c) Geochemical evolution of the crust and mantle. Rev Geophys Space Phys 21:1347 – 1358

DePaolo DJ (1985) Isotopic studies of processes in mafic magma chambers. I. The Kiglapait intrusion, Labrador. J Petrol 26:925 – 951

DePaolo DJ (1988) Age dependence of the composition of continental crust: evidence from Nd isotopic variations in granitic rocks. Earth Planet Sci Lett (in press)

DePaolo DJ, Johnson RW (1979) Magma genesis in the New Britain island arc: constraints from Nd and Sr isotopes and trace element patterns. Contrib Mineral Petrol 70:367 – 379

DePaolo DJ, Wasserburg GJ (1976a) Nd isotopic variations and petrogenetic models. Geophys Res Lett 3:249 – 252

DePaolo DJ, Wasserburg GJ (1976b) Inferences about magma sources and mantle structure from variations of ^{143}Nd/^{144}Nd. Geophys Res Lett 3:743 – 746

DePaolo DJ, Wasserburg GJ (1977) The sources of island arcs as indicated by Nd and Sr isotopic studies. Geophys Res Lett 4:465 – 468

DePaolo DJ, Wasserburg GJ (1979a) Sm-Nd age of the Stillwater complex and the mantle evolution curve for neodymium. Geochim Cosmochim Acta 43:99 – 1008

DePaolo DJ, Wasserburg GJ (1979b) Nd isotopes in flood basalts from the Siberian Platform and inferences about their mantle sources. Proc Natl Acad Sci USA 76(7):3056 – 3060

DePaolo DJ, Wasserburg GJ (1979c) Petrogenetic mixing models and Nd-Sr isotopic patterns. Geochim Cosmochim Acta 43:615 – 627

DePaolo DJ, Manton WI, Grew ES, Halpern M (1982) Sm-Nd, Rb-Sr and U-Th-Pb systematics of granulite facies rocks from Fyfe Hills, Enderby Land, Antarctica. Nature 298:614 – 618

Diskinson WR (1970) Relations of andesites, granites and derivative sandstones to arc-trench tectonics. Rev Geophys Space Phys 8:813 – 860

Doe BR (1970) Lead isotopes. Springer, Berlin Heidelberg New York, 221 pp

Doe BR, Zartman RE (1979) Plumbotectonics, the Phanerozoic. In: Barnes HL (ed) Geochemistry of hydrothermal ore deposits. Wiley, New York, pp 22 – 70

Dosso L, Murthy VR (1980) A Nd isotopic study of the Kerguelen Islands: inferences on enriched oceanic mantle sources. Earth Planet Sci Lett 48:268 – 276

Drake MJ (1976) Evolution of major mineral compositions and trace element abundances during fractional crystallization of a model lunar composition. Geochim Cosmochim Acta 40: 401 – 411

Drake MJ, Weill DF (1975) Partition of Sr, Ba, Ca, Y, Eu^{2+}, Eu^{3+} and other REE between plagioclase feldspar and magmatic liquid: an experimental study. Geochim Cosmochim Acta 39:689 – 712

Dupre B, Allegre CJ (1980) Pb-Sr-Nd isotopic correlation and the chemistry of the North Atlantic mantle. Nature 286:17

Dupre B, Chauvel C, Arndt NT (1984) Pb and Nd isotopic study of two Archean komatiite flows from Alexo, Ontario. Geochim Cosmochim Acta 48:1965 – 1972

Edmond JM, Measures C, McDuff RE, Chan LH, Collier R, Grant G, Gordon LI, Corliss JB (1982) Ridge crest hydrothermal activity and the balances of the major and minor elements in the ocean: the Galapagos data. Earth Planet Sci Lett 57:191 – 210

Elthon D (1979) High magnesia liquids as the parental magma for ocean floor basalts. Nature 278:514 – 518

Eugster O, Tera F, Burnett DS, Wasserburg GJ (1970) The isotopic composition of gadolinium and neutron capture effects in some meteorites. J Geophys Res 75:2753 – 2768

Farmer GL, DePaolo DJ (1983) Origin of Mesozoic and Tertiary granite in the western US and implications for pre-Mesozoic crustal struture. I. Nd and Sr isotopic studies in the geocline of the northern Great Basin. J Geophys Res 88:3379 – 3401

Farmer GL, DePaolo DJ (1984) Origin of Mesozoic and Tertiary granite in the western US and implications for pre-Mesozoic crustal structure. II. Nd and Sr isotopic studies of unmineralized and Cu- and Mo-mineralized granite in the Precambrian craton. J Geophys Res 89: 10141 – 10160

Faure G (1977) Principles of isotope geology. Wiley, New York, 464 p

Faure G, Hurley PM (1963) The isotopic composition of strontium in oceanic and continental basalt: application to the origin of igneous rocks. J Petrol 4:31 – 50

Faure G, Powell JL (1972) Strontium isotope geology. Springer, Berlin Heidelberg New York, 188 p

Fletcher RC, Hofmann AW (1974) Simple models of diffusion and combined diffusion-infiltration metasomatism. In: Hofmann AW et al. (eds) Geochemical transport and kinetics. Carnegie Inst Wash Publ 634:243 – 259

Francis PW, Moorbath S, Thorpe RS (1977) Strontium isotope data for recent andesites in Ecuador and north Chile. Earth Planet Sci Lett 37:197

Francis PW, Thorpe RS, Moorbath S, Kretzschmar GA, Hammill M (1980) Strontium isotope evidence for crustal contamination of calc-alkaline volcanic rocks from Cerro Galan, northwest Argentina. Earth Planet Sci Lett 48:257

Frey FA (1982) Rare earth element abundances in upper mantle rocks. In: Henderson P (ed) Rare element geochemistry, chapter 5. Elsevier, Amsterdam

Frey FA, Haskin LA, Haskin MA (1971) Rare earth abundances in some ultramafic rocks. J Geophys Res 76:2057 – 2070

Frey FA, Gren DH, Roy SD (1978) Integrated models of basalt petrogenesis: a study of quartz tholeiites to olivine melilitites from southeastern Australia utilizing geochemical and experimental petrological data. J Petrol 19:463 – 513

Fujii T, Bougault H (1983) Melting relations of a magnesian abyssal tholeiite and the origin of MORB's. Earth Planet Sci Lett 62:283 – 295

Ganapathy RM, Anders E (1974) Bulk composition of the moon and earth estimated from meteorites. Proc Lunar Planet Sci Conf 5th, pp 1181 – 1206

Gast PW (1960) Limitations on the composition of the upper mantle. J Geophys Res 65:1287 – 1297

Gast PW (1968) Trace element fractionation and the origin of tholeiitic and alkaline magma types. Geochim Cosmochim Acta 32:1057 – 1086

Gast PW, Tilton GR, Hedge CE (1964) Isotopic composition of lead and strontium from Ascension and Gough Islands. Science 145:1181 – 1185

Geiss J, Eberhardt P, Grogler N, Guggisberg S, Maurer P, Stettler A (1977) Absolute time scale of lunar mare formation and filling. Philos Trans R Soc Lond A Math Phys Sci 285:151 – 158

General Electric Company (1972) Chart of the Nuclides. San Jose, California, 48 p

Gill J (1981) Orogenic andesites and plate tectonics. Springer, Berlin Heidelberg New York, 390 p

Goldich SS (1968) Geochronology in the Lake Superior region. Can J Earth Sci 5:715 – 724

Green DH (1971) Composition of basaltic magmas as indicators of conditions of origin: applications to oceanic volcanism. Philos Trans R Soc Lond A Math Phys Sci 268:707 – 725

Green DH (1975) Genesis of Archean peridotitic magmas and constraints on Archean geothermal gradients and tectonics. Geology (Boulder) 3:15 – 18

Green DH, Ringwood AE (1967) The genesis of basaltic magmas. Contrib Mineral Petrol 15:103 – 190

Gregory RT, Taylor HP (1981) An oxygen isotope profile in a section of Cretaceous oceanic crust, Semail ophiolite, Oman: evidence of $\delta^{18}O$-buffering of the oceans by deep (>5 km) seawater-hydrothermal circulation at mid-ocean ridges. J Geophys Res 86:2737 – 2755

Griffin WL, Brueckner HK (1978) Caledonian Sm-Nd ages and a crustal origin for Norwegian eclogites. Nature 285:319 – 321

Grossman L, Larimer JW (1974) Early chemical history of the solar system. Rev Geophys Space Phys 12:71 – 101

Gupta MC, McFarlane RD (1970) The natural alpha radioactivity of samarium. J Inorg Nucl Chem 32:3425 – 3432

Hager BH, O'Connell RJ (1981) A simple global model of plate dynamics and mantle convection. J Geophys Res 86:4843

Hamelin B, Dupre B, Allegre CJ (1986) Pb-Nd-Sr isotopic data of Indian Ocean ridges: new evidence of large scale mapping of mantle heterogeneities. Earth Planet Sci Lett 76:288 – 296

Hamet J, Nakamura DM, Unruh DM, Tatsumoto M (1978) Origin and history of the adcumulate eucrite, Moama as inferred from REE abundances, Sm-Nd and U-Pb systematics. Proc. Lunar Planet Sci Conf 9th, pp 1115 – 1136

Hamilton PJ, O'Nions RK, Evensen NM (1977) Sm-Nd dating of Archean basic and ultrabasic volcanics. Earth Planet Sci Lett 36:263 – 268

Hamilton PJ, O'Nions RK, Evensen NM, Bridgewater D, Allart JH (1978) Sm-Nd isotopic investigations of Isua supracrustals and implications for mantle evolution. Nature 272:41 – 43

Hamilton PJ, Evensen NM, O'Nions RK, Smith HS, Erlank AJ (1979a) Sm-Nd dating of Onverwacht volcanics, southern Africa. Nature 279:298 – 301

Hamilton PJ, Evensen NM, O'Nions RK, Tarney J (1979b) Sm-Nd systematics of Lewisian gneisses: implications for the origin of granulites. Nature 277:25 – 29

Hamilton PJ, O'Nions RK, Pankhurst RJ (1980) Isotopic evidence for the provenance of some Caledonian granites. Nature 287:279 – 284

Hanson GN (1980) Rare earth elements in petrogenetic studies of igneous systems. Ann Rev Earth Planet Sci 8:371 – 406

Harris PG (1974) Origin of alkaline magmas as a result of anatexis. In: Sorensen H (ed) The alkaline rocks. Wiley, New York, 427 p

Hart R, Dymond J, Hogan L (1979) Preferential formation of the atmosphere-sialic crust system from the upper mantle. Nature 278:156 – 158

Hart SR (1964) The petrology and isotopic-mineral age relations of a contact zone in the Front Range, Colorado. J Geol 72:493 – 525

Hart SR (1984) A large-scale isotope anomaly in the Southern Hemisphere mantle. Nature 309:753 – 757

Hart SR, Allegre CJ (1980) Trace element constraints on magma genesis. In: Hargraves RB (ed) Physics of magmatic processes. Princeton University Press, pp 121 – 159

Hart SR, Brooks C (1980) Sources of terrestrial basalts: an isotopic viewpoint. In: Basaltic volcanism of the terrestrial planets, chapter 7. Lunar&Planetary Institute, Houston

Hart SR, Staudigel H (1982) The control of alkalies and uranium in seawater by ocean crust alteration. Earth Planet Sci Lett 58:202 – 212

Haskin LA, Paster TP (1979) Geochemistry and mineralogy of the rare earths. In: Gschneidner KA (ed) Handbook on the physics and chemistry of the rare earths, Chap. 21. North-Holland, Amsterdam

Haskin LA, Frey FA, Schmitt RA, Smith RH (1966) Meteoritic, solar, and terrestrial rare-earth distributions. Phys Chem Earth 7:167 – 321

Hawkesworth CJ (1979) ^{143}Nd/^{144}Nd, ^{87}Sr/^{86}Sr and trace element characteristics of magmas along destructive plate margins. In: Atherton MP, Tarney J (eds) Origin of granite batholiths. Shiva, Kent, Great Britain, pp 76 – 89

Hawkesworth CJ (1982) Isotope characteristics of magmas erupted along destructive plate margins. In: Thorpe RS (ed) Andesites: orogenic andesites and related rocks. Wiley, New York, pp 549 – 571

Hawkesworth CJ, Powell M (1980) Magma genesis in the Lesser Antilles island arc. Earth Planet Sci Lett 51:297 – 308

Hawkesworth CJ, Vollmer R (1979) Crustal contamination versus enriched mantle; ^{143}Nd/^{144}Nd and ^{87}Sr/^{86}Sr evidence from the Italian volcanics. Contrib Mineral Petrol 69:151 – 165

Hawkesworth CJ, O'Nions RK, Pankhurst RJ, Hamilton PJ, Evensen NM (1977) A geochemical study of island arc and back-arc tholeiites from the Scotia Sea. Earth Planet Sci Lett 36:253 – 262

Hawkesworth CJ, Norry MJ, Roddick JC, Vollmer R (1978) The significance of trace element modeling calculations for the evolution of Sr and Nd isotopes in the mantle. US Geol Surv Open-File Rep 78-701:162 – 164

Hawkesworth CJ, Norry MJ, Roddick JC, Vollmer R (1979a) ^{143}Nd/^{144}Nd and ^{87}Sr/^{86}Sr ratios from the Azores and their significance in LIL-enriched mantle. Nature 280:28 – 31

Hawkesworth CJ, Norry MJ, Roddick JC, Baker PE, Francis PW, Thorpe RS (1979b) ^{143}Nd/^{144}Nd, ^{87}Sr/^{86}Sr, and incompatible element abundances in calc-alkaline andesites and plateau lavas from South America. Earth Planet Sci Lett 42:45 – 57

Hawkesworth CJ, O'Nions RK, Arculus RJ (1979c) Nd and Sr isotope geochemistry of island arc volcanics, Grenada, Lesser Antilles. Earth Planet Sci Lett 45:237 – 248

Hawkesworth CJ, Hammill M, Gledhill AR, Calsteren P van, Rogers G (1982a) Isotope and trace element evidence for late-stage intracrustal melting in the high Andes. Earth Planet Sci Lett 58:240 – 254

Hawkesworth CJ, Erlank AJ, Menzies MA, Calsteren P van (1982b) Late Proterozoic lithosphere beneath southern Africa: isotope and trace element evidence from Karroo lavas and crustal and mantle xenoliths. Abs Fifth Intern Conf Geochronology, Cosmochronology, and Isotope Geology, pp 135 – 136

Hawkesworth CJ, Rogers NW, Calsteren PWC van, Menzies MA (1984) Mantle enrichment processes. Nature 311:331 – 335

Hedge CE (1966) Variations in radiogenic strontium found in volcanic rocks. J Geophys Res 71:6119 – 6126

Heier KS (1965) Metamorphism and the chemical differentiation of the crust. Geol Foren Stockh Forh 87:249 – 256

Hofmann AW, Hart SR (1978) An assessment of local and regional isotopic equilibrium in the mantle. Earth Planet Sci Lett 38:44 – 62

Hofmann AW, Magaritz M (1977) Equilibrium and mixing in a partially molten mantle. In: Dick HB (ed) Magma genesis, Bull 96, Oregon Department of Geology and Mineral Industries

Hofmann AW, White WM (1982) Mantle plumes from ancient oceanic crust. Earth Planet Sci Lett 57:421 – 436

Hofmeister AM (1983) Effect of a Hadean terrestrial magma ocean on crust and mantle evolution. J Geophys Res 88:4963 – 4983

Högdahl OT, Melson S, Bowen V (1968) Neutron activation analysis of lanthanide elements in seawater. Adv Chem Ser 73:308 – 325

Hurley PM (1968) Absolute abundance and distribution of Rb, K, and Sr in the earth. Geochim Cosmochim Acta 32:273 – 283

Hurley PM, Rand R (1969) Pre-drift continental nucleii. Science 164:1229 – 1242

Inghram MG, Chupka WA (1953) Surface ionization source using multiple filaments. Rev Sci Instrum 24:518 – 520

Irving AJ (1978) A review of experimental studies of crystal/liquid trace element partitioning. Geochim Cosmochim Acta 42:743 – 770

Jacobsen SB, Wasserburg GJ (1978a) The interpretation of Nd, Sr, and Pb isotope data from Archean migmatites in Lofoten-Vesteraalen, Norway. Earth Planet Sci Lett 41:245 – 253

Jacobsen SB, Wasserburg GJ (1978b) Nd and Sr isotopic study of the Permian Oslo Rift. US Geol Surv Open File Rep 78-701:194 – 196

Jacobsen SB, Wasserburg GJ (1979a) Nd and Sr isotopic study of the Bay of Islands ophiolite complex and the evolution of the source of mid-ocean ridge basalts. J Geophys Res 84:7429 – 7445

Jacobsen SB, Wasserburg GJ (1979b) The mean age of mantle and crustal reservoirs. J Geophys Res 84:7411

Jacobsen SB, Wasserburg GJ (1980a) Sm-Nd evolution of chondrites. Earth Planet Sci Lett 50:139 – 155

Jacobsen SB, Wasserburg GJ (1980b) A two-reservoir recycling model for mantle-crust evolution. Proc Nat Acad Sci USA 77:6298 – 6302

Jacobsen SB, Wasserburg GJ (1984) Sm-Nd isotopic evolution of chondrites and achondrites, II. Earth Planet Sci Lett 67:137 – 150

Jacques AL, Green DH (1980) Anhydrous melting of peridotite at 0 – 15 kb pressure and the genesis of tholeiitic basalts. Contrib Mineral Petrol 73:287 – 310

Jahn B-M, Bernard-Griffiths J, Charlot R, Cornichet J, Vidal F (1980) Nd and Sr isotopic compositions and REE abundances of Cretaceous MORB (Holes 417D and 418A, Legs 51, 52 and 53). Earth Planet Sci Lett 48:171 – 184

James DE (1982) A combined O, Sr, Nd, and Pb isotopic and trace element study of crustal contamination in central Andean lavas, I. Local geochemical variations. Earth Planet Sci Lett 57:47 – 62

James DE, Brooks C, Cuyubamba A (1976) Andean Cenozoic volcanism: magma genesis in the light of strontium isotope composition and trace element geochemistry. Geol Soc Am Bull 87:592

Jeanloz R, Richter FM (1979) Convection, composition, and the thermal state of the lower mantle. J Geophys Res 84:5497

Johnson LR (1967) Array measurements of P velocities in the upper mantle. J Geophys Res 72:6309 – 6325

Johnson LR (1969) Array measurements of P-velocities in the lower mantle. Bull Seism Soc Am 59:973 – 1011

Jones AP, Smith JV, Dawson JB (1982) Mantle metasomatism in 14 veined peridotites from Bultfontein Mine, South Africa. J Geol 90:435 – 453

Kaneoka I, Takaoka N (1980) Rare gas isotopes in Hawaiian ultramafic nodules and volcanic rocks: constraints on genetic relationships. Science 208:1366 – 1368

Kaula WM (1979) Thermal evolution of the earth and moon growing by planetesimal impacts. J Geophys Res 84:999 – 1008

Kay RW (1980) Implications of a melting-mixing model for element recycling in the crust-mantle system. J Geol 88:497 – 522

Kay RW, Gast PW (1973) The rare-earth content and origin of alkali-rich basalts. J Geol 81:653 – 682

Kay RW, Hubbard NJ (1978) Trace elements in ocean ridge basalts. Earth Planet Sci Lett 38:95 – 116

Kay RW, Rubenstone JL, Kay SM (1986) Aleutian terranes from Nd isotopes. Nature 322:605 – 609

Kistler RW, Peterman ZE (1973) Variations in Sr, Rb, K, Na and initial $^{87}Sr/^{86}Sr$ in Mesozoic granitic rocks and intruded wall rocks in central California. Geol Soc Am Bull 84:3489 – 3512

Kramers JD, Smith CB, Lock NP, Harmon RS, Boyd FR (1981) Can kimberlites be generated from ordinary mantle? Nature 291:53 – 56

Krogh TE (1982a) Improved accuracy of U-Pb zircon dating by selection of more concordant fractions using a high gradient magnetic separation technique. Geochim Cosmochim Acta 46:631 – 636

Krogh TE (1982b) Improved accuracy of U-Pb zircon ages by the creation of more concordant systems using an air abrasion technique. Geochim Cosmochim Acta 46:637 – 650

Kurz MD, Jenkins WJ (1981) The distribution of helium in oceanic basalt glasses. Earth Planet Sci Lett 53:41 – 54

Kurz MD, Jenkins WJ, Schilling J-G (1982a) Helium isotopic variations in the mantle beneath the central North Atlantic Ocean. Earth Planet Sci Lett 58:1 – 14

Kurz MD, Jenkins WJ, Hart SR (1982b) Helium isotopic systematics in oceanic islands and mantle heterogeneity. Nature 297:43

Kyser TK, Rison W (1982) Systematics of rare gas isotopes in basic lavas and ultramafic xenoliths. J Geophys Res 87:5611

Langmuir CH, Vocke RD, Hanson GN (1978) A general mixing equation with application to Icelandic basalts. Earth Planet Sci Lett 37:380 – 392

Lanphere MA, Wasserburg GJF, Albee AL, Tilton GR (1963) Redistribution of strontium and rubidium isotopes during metamorphism, World Beater Complex, Panamint Range, California. In: Craig H, Miller SL, Wasserburg GJ (eds) Isotopic and cosmic chemistry. North-Holland Publishing Company, Amsterdam, pp 269 – 320

Lewis JS (1971) Consequences of the presence of sulfur in the core of the earth. Earth Planet Sci Lett 11:130 – 134

Liew TC, McCulloch MT (1985) Genesis of granitoid batholiths of Peninsular Malaysia and implications for models of crustal evolution: evidence from a Nd-Sr and U-Pb zircon study. Geochim Cosmochim Acta 49:587 – 600

Longhi J (1980) A model of early lunar differentiation. Proc Lunar Planet Sci 11th, pp 289 – 315

Lugmair GW (1974) Sm-Nd ages: a new dating method. Meteoritics 9:369

Lugmair GW, Marti K (1977) Sm-Nd-Pu timepieces in the Angra Dos Reis meteorite. Earth Planet Sci Lett 35:273 – 284

Lugmair GW, Marti K (1978) Lunar initial $^{143}Nd/^{144}Nd$; differential evolution of the lunar crust and mantle. Earth Planet Sci Lett 39:349 – 357

Lugmair GW, Scheinin NB, Marti K (1975a) Sm-Nd age of Apollo 17 basalt 75075: evidence for early differentiation of the lunar exterior. Proc Lunar Planet Sci Conf 6th, pp 1419 – 1429

Lugmair GW, Scheinin NB, Marti K (1975b) Search for extinct ^{146}Sm: 1. The isotopic abundance of ^{142}Nd in the Juvinas meteorite. Earth Planet Sci Lett 27:79 – 84

Lugmair GW, Scheinin NB, Marti K (1976) History and genesis of lunar troctolite 76535 or: How old is old? Proc Lunar Planet Sci Conf 7th, pp 2009 – 2033

Lugmair GW, Marti K, Scheinin NB (1978) Incomplete mixing of products from r-, p-, and s-process nucleosynthesis: Sm-Nd systematics in Allende inclusion EK1-4-1. Lunar Planet IX:672 – 674

Lunatic Asylum (1978) Petrology, chemistry, age and irradiation history of Luna 24 samples. In: Merrill RB, Papike JJ (eds) Mare crisium: the view from Luna 24. Pergamon, New York, pp 657-678

Lupton JE (1983) Terrestrial inert gases: isotope tracer studies and clues to primordial constituents in the mantle. Ann Rev Earth Planet Sci 11:371 – 414

Maaloe S (1981) Magma accumulation in the ascending mantle. J Geol Soc (Lond) 138:223 – 236

MacDougall JD, Lugmair GW (1985) Extreme isotopic homogeneity among basalts from the southern East Pacific Rise: mantle or mixing effect? Nature 313:209 – 211

Mahoney J, Mac Dougall JD, Lugmair GW, Murali AV, Sankar Das M, Gopalan K (1982) Origin of the Deccan Trap flows at Mahabaleshwar inferred from Nd and Sr isotopic and chemical evidence. Earth Planet Sci Lett 60:47 – 60

Mason B, Moore CB (1982) Principles of geochemistry, 4th edn. Wiley, New York

Mattinson JM (1975) Early Paleozoic ophiolite complexes of Newfoundland: isotopic ages of zircons. Geology (Boulder) 3:181 – 183

Mattinson JM (1976) Ages of zircons from the Bay of Islands ophiolite complex, western Newfoundland. Geology (Boulder) 393 – 394

McCulloch MT, Chappell BW (1982) Nd isotopic characteristics of S- and I-type granites. Earth Planet Sci Lett 58:51 – 64

McCulloch MT, Compston W (1981) Sm-Nd age of the Kambalda and Kanowna greenstones and heterogeneity in the Archean mantle. Nature 294:322

McCulloch MT, Wasserburg GJ (1978a) Ba and Nd isotopic anomalies in the Allende meteorite. Astrophys J Lett 220:L15 – 19

McCulloch MT, Wasserburg GJ (1978b) More anomalies from the Allende meteorite: samarium. Geophys Res Lett 5:599 – 602

McCulloch MT, Wasserburg GJ (1978c) Sm-Nd and Rb-Sr chronology of continental crust formation. Science 200:1003 – 1011

McCulloch MT, Gregory RT, Wasserburg GJ, Taylor HP (1980) A neodymium, strontium, and oxygen isotopic study of the Cretaceous Semail ophiolite and implications for the petrogenesis and seawater-hydrothermal alteration of oceanic crust. Earth Planet Sci Lett 46:201 – 211

McCulloch MT, Gregory RT, Wasserburg GJ, Taylor HP (1981) Sm-Nd, Rb-Sr and $^{18}O/^{16}O$ isotopic systematics in an oceanic crustal section: evidence from the Semail ophiolite. J Geophys Res 86:2721

McCulloch MT, Jacques AL, Nelson DR, Lewis JD (1983) Nd and Sr isotopes in kimberlites and lamproites from western Australia: an enriched mantle origin. Nature 302:400 – 403

McDougall I (1966) Precision methods of potassium-argon age determination on young rocks. In: Methods and techniques in geophysics, vol 2. Interscience, New York, pp 279 – 304

McKay GA (1986) Crystal/liquid partitioning of REE in basaltic systems: extreme fractionation of REE in olivine. Geochim Cosmochim Acta 50:69 – 79

McLennan SM, Taylor SR (1984) Archean sedimentary rocks and their relation to the composition of the Archean continental crust. In: Kroner A, Hanson GN, Goodwin AM (eds) Archean geochemistry. Springer, Berlin Heidelberg New York, pp 47 – 72

Menzies M, Murthy VR (1979) Nd and Sr isotope geochemistry of hydrous mantle nodules and their host alkali basalts: implications for local heterogeneities in metasomatically-veined mantle. Earth Planet Sci Lett 46:323 – 334

Menzies M, Murthy VR (1980a) Enriched mantle: Nd and Sr isotopes in diopsides from kimberlite nodules. Nature 283:634

Menzies M, Murthy VR (1980b) Mantle metasomatism as a precursor to the genesis of alkaline magmas – isotopic evidence. Am J Sci 280-A:622 – 638

Menzies M, Seyfried D, Blanchard D (1979) Experimental evidence of rare earth element mobility in greenstones. Nature 282:398

Menzies M, Leeman WP, Hawkesworth CJ (1983) Isotope geochemistry of Cenozoic volcanic rocks reveals mantle heterogeneity below western USA. Nature 303:205 – 207

Michard-Vitrac A, Lancelot J, Allegre CJ, Moorbath S (1977) U-Pb ages on single zircons from the early Precambrian rocks of West Greenland and the Minnesota River valley. Earth Planet Sci Lett 35:449 – 453

Miller RG, O'Nions RK (1984) The provenance and crustal residence ages of British sediments in relation to paleogeographic reconstructions. Earth Planet Sci Lett 68:459 – 470

Minear JW, Fletcher CR (1978) Crystallization of a lunar magma ocean. Proc Lunar Planet Sci Conf 9th, pp 263 – 283

Minster JB, Jordan TH (1978) Present-day plate motions. J Geophys Res 83:5331 – 5354

Minster J-F, Birck J-L, Allegre CJ (1982) Absolute age of formation of chondrites studied by the ^{87}Rb-^{87}Sr method. Nature 300:414 – 419

Moorbath S, Welke H, Gale NH (1969) The significance of lead isotope studies in ancient, high-grade metamorphic basement complexes, as typified by the Lewisian rocks of northwest Scotland. Earth Planet Sci Lett 6:245 – 256

Morgan WJ (1971) Convection plumes in the lower mantle. Nature 230:42 – 43

Morris JD, Hart SR (1983) Isotopic and incompatible element constraints on the genesis of island arc volcanics from Cold Bay and Amak Island, Aleutians, and implications for mantle structure. Geochim Cosmochim Acta 47:2015 – 2030

Morse SA (1969) Geology of the Kiglapait Intrusion, Labrador. Mem Geol Soc Am 112

Morse SA (1980) Basalts and phase diagrams. Springer, Berlin Heidelberg New York, 493 p

Muehlburger WR (1980) The shape of North America during the Precambrian. In: Continental Tectonics. Washington DC, National Academy of Sciences, pp 175 – 183

Murthy VR, Hall H (1970) The chemical composition of the earth's core: possibility of sulfur in the core. Phys Earth Planet Inter 2:276 – 282

Musselwhite DM, DePaolo DJ, McCurry MJ (1987) The evolution of silicic magma systems: isotopic and chemical evidence from the Woods Mountains Volcanic Center, eastern California. Contrib Mineral Petrol (in press)

Mysen BO, Kushiro I (1977) Compositional variations of coexisting phases with degree of melting of peridotite in the upper mantle. Am Miner 62:843 – 865

Nakamura N, Tatsumoto M, Nunes PD, Unruh DM, Schwab AP, Wildeman TR (1976) 4.4 b.y.-old clast in Boulder 7, Apollo 17: a comprehensive chronological study by U-Pb, Rb-Sr and Sm-Nd methods. Proc Lunar Planet Sci Conf 7th, 2:2309 – 2333

Naudet R (1978) Les reacteur d'Oklo: Cinq ans d'exploration du site. In: Les Reacteur De Fission Naturel. International Atomic Energy Agency, Vienna, pp 3 – 18

Nelson BK, DePaolo DJ (1984) Origin of 1700 Myr greenstone successions in southwestern North America and the isotopic evolution of Proterozoic mantle. Nature 311:143 – 146

Nelson BK, DePaolo DJ (1985) Rapid production of continental crust 1.7 – 1.9 b.y. ago: Nd and Sr isotopic evidence from the basement of the North American midcontinent. Geol Soc Am Bull 96:746 – 754

Nelson BK, DePaolo DJ (1988) Application of Sm-Nd and Rb-Sr isotope systematics to studies of provenance and basin analysis. J Sediment Petrol 58:348 – 357

Niazi M, Anderson DA (1965) Upper mantle structure of western North America from apparent velocities of P waves. J Geophys Res 70:4633 – 4640

Nicholls J, Carmichael ISE, Stormer (1971) Silica activity and P-total in igneous rocks. Contrib Mineral Petrol 33:1 – 20

Nier AO (1940) A mass spectrometer for routine isotope abundance measurements. Rev Sci Instrum 11:212 – 216

Nier AO (1947) A mass spectrometer for isotope and gas analysis. Rev Sci Instrum 18:398 – 411

Nier AO (1950) A redetermination of the relative abundances of the isotopes of carbon, nitrogen, oxygen, argon, and potassium. Phys Rev 77:789 – 793

Nohda S, Wasserburg GJ (1981) Nd and Sr isotopic study of volcanic rocks from Japan. Earth Planet Sci Lett 52:264 – 276

Nunes PD (1981) The age of the Stillwater complex – a comparison of U-Pb zircon and Sm-Nd isochron systematics. Geochim Cosmochim Acta 45:1961 – 1963

Nyquist LE, Shih C-Y, Wooden JL, Bansal BM, Wiesman H (1979) The Sr and Nd isotopic record of Apollo 12 basalts: implications for lunar geochemical evolution. Geochim Cosmochim Acta 10:77 – 114

O'Hara MJ (1968) Are ocean floor basalts primary magma? Nature 220:683 – 686

O'Hara MJ (1985) The importance of the "shape" of the melting regime during partial melting of the mantle. Nature 314:58 – 62

O'Hara MJ, Mathews RE (1981) Geochemical evolution in an advancing, periodically replenished, periodically tapped, continuously fractionating magma chamber. J Geol Soc (Lond) A 138:237 – 277

O'Nions RK (1984) Isotopic abundances relevant to the identification of magma sources. Philos Trans R Soc Lond Math Phys Sci 310:591 – 603

O'Nions RK, Hamilton PJ, Evenson NM (1977) Variations in ^{143}Nd/^{144}Nd and ^{87}Sr/^{86}Sr in oceanic basalts. Earth Planet Sci Lett 34:13 – 22

O'Nions RK, Carter SR, Cohen RS, Evensen NM, Hamilton PJ (1978) Pb, Nd and Sr isotopes in oceanic ferromanganese deposits and ocean floor basalts. Nature 273:435 – 438

O'Nions RK, Carter SR, Evensen NM, Hamilton PJ (1979a) Geochemical and cosmochemical applications of Nd isotope analysis. Ann Rev Earth Planet Sci 7:11 – 38

O'Nions RK, Evensen NM, Hamilton PJ (1979b) Geochemical modeling of mantle differentiation and crustal growth. J Geophys Res 84:6091

O'Nions RK, Hamilton PJ, Hooker PJ (1983) A Nd isotope investigation of sediments related to crustal development in the British Isles. Earth Planet Sci Lett 63:229 – 240

Olson P, Yuen DA (1982) Thermochemical plumes and mantle phase transitions. J Geophys Res 87:3993 – 4002

Papanastassiou DA, Wasserburg GJ (1969) Initial strontium isotopic abundances and the resolution of small time differences in the formation of planetary objects. Earth Planet Sci Lett 5:361 – 376

Papanastassiou DA, DePaolo DJ, Wasserburg GJ (1977) Rb-Sr and Sm-Nd chronology and genealogy of mare basalts from the Sea of Tranquility. Proc Lunar Planet Sci Conf 8th, pp 1639 – 1672

Patchett PJ (1982) Hf isotopes and mantle evolution. Abs Fifth Intern Conf Geochronology, Cosmochronology and Isotope Geology, Nikko, Japan, pp 305 – 306

Patchett PJ (1983) Importance of Lu-Hf isotopic system in studies of planetary chronology and chemical evolution. Geochim Cosmochim Acta 47:81 – 91

Patchett PJ, Kuovo O, Hedge CE, Tatsumoto M (1981) Evolution of continental crust and mantle heterogeneity. Contrib Mineral Petrol 78:279 – 297

Patchett PJ, Tatsumoto M (1980a) Hafnium isotope variations in oceanic basalts. Geophys Res Lett 7:1077 – 1080

Patchett PJ, Tatsumoto M (1980b) A routine high-precision method for Lu-Hf isotope geochemistry and chronology. Contrib Mineral Petrol 75:263 – 267

Patterson CC (1956) Age of meteorites and the earth. Geochim Cosmochim Acta 10:230 – 237

Patterson CC (1964) Characteristics of lead isotope evolution on a continental scale in the earth. In: Craig H (ed) Isotopic and cosmic chemistry. North Holland, Amsterdam, pp 244 – 268, 553

Peppard DF (1961) Separation of the rare earths by liquid-liquid extraction. In: Spedding FH, Daane AH (eds) The rare earths. Wiley, New York, pp 38 – 54

Perry FV, Baldridge WS, DePaolo DJ (1987) Role of asthenosphere and lithosphere in the genesis of Late Cenozoic basaltic rocks from the Rio Grande Rift and adjacent regions of the southwestern United States. J Geophys Res 92:9193 – 9213

Piepgras DJ, Wasserburg GJ (1980) Neodymium isotopic variations in seawater. Earth Planet Sci Lett 50:128 – 138

Piepgras DJ, Wasserburg GJ, Dasch EJ (1979) The isotopic composition of Nd in different ocean masses. Earth Planet Sci Lett 45:223 – 236

Powell JE (1961) Separation of the rare earths by ion exchange. In: Spedding FH, Daane AH (eds) The rare earths. Wiley, New York, pp 55 – 73

Rankama K (1954) Isotope geology. Pergamon, London

Rankama K (1963) Progress in isotope geology. Interscience, New York

Richard P, Allegre CJ (1980) Neodymium and strontium isotope study of ophiolite and orogenic lherzolite petrogenesis. Earth Planet Sci Lett 47:65 – 74

Richard P, Shimizu N, Allegre CJ (1976) $^{143}Nd/^{146}Nd$, a natural tracer. An application to oceanic basalts. Earth Planet Sci Lett 31:269 – 278

Richardson SH, Erlank AJ, Duncan AR, Reid DL (1982) Correlated Nd, Sr, and Pb isotope variation in Walvis Ridge basalts and implications for the evolution of their mantle source. Earth Planet Sci Lett 59:327 – 342

Richter FM (1986) Simple models for trace element fractionation during melt segregation. Earth Planet Sci Lett 77:333 – 344

Richter FM, Johnson CE (1974) Stability of a chemically layered mantle. J Geophys Res 79:1635 – 1639

Richter FM, McKenzie DP (1979) On some consequences and possible causes of layered mantle convection. J Geophys Res 86:6133

Richter FM, McKenzie DP (1984) Dynamical models of melt segregation from a deformable matrix. J Geol 92:729 – 740

Richter FM, Ribe NM (1979) On the importance of advection in determining the local isotopic composition of the mantle. Earth Planet Sci Lett 43:212 – 222

Richter FM, Daly SF, Nataf H-C (1982) A parameterized model for the evolution of isotopic heterogeneities in a convecting system. Earth Planet Sci Lett 60:178 – 194

Ringwood AE (1975) Composition and petrology of the earth's mantle. McGraw-Hill, New York

Russ GP (1974) Neutron stratigraphy in the lunar regolith. Ph D Thesis, California Institute of Technology

Russ GP, Burnett DS, Lingenfelter RE, Wasserburg GJ (1971) Neutron capture on 149-Sm in lunar samples. Earth Planet Sci Lett 13:53 – 60

Russell RD, Farquhar RM (1960) Lead isotopes in geology. Interscience, New York

Russell WA, Papanastassiou DA, Tombrello TA (1977) Ca isotope fractionation in the earth and other solar system materials. Geochim Cosmochim Acta 42:1075 – 1090

Schilling J-G (1973) Icelandic mantle plume: geochemical evidence along the Reykjanes Ridge. Nature 242:565 – 571

Schilling J-G, Winchester JW (1967) Rare earth fractionation and magmatic processes. In: Runcorn SK (ed) Mantles of the earth and terrestrial planets. Interscience, London, pp 267 – 283

Schilling J-G, Winchester JW (1969) Rare earth contribution to the origin of Hawaiian lavas. Contrib Mineral Petrol 23:27 – 37

Schnetzler CC, Philpotts JA (1970) Partition coefficients of rare earth elements between igneous matrix material and rock forming mineral phenocrysts. Geochim Cosmochim Acta 34:331 – 340

Schubert G, Spohn T (1981) Two-layer mantle convection and the depletion of radioactive elements in the lower mantle. Geophys Res Lett 8:951 – 954

Semken SC (1984) A neodymium and strontium isotopic study of late Cenozoic basaltic volcanism in the southwestern Basin and Range province. MS Thesis, University of California, Los Angeles, 68 p

Semken SC, DePaolo DJ (1983) Sm-Nd and Rb-Sr constraints on Cenozoic basalt petrogenesis and mantle heterogeneity in the SW Great Basin. EOS 64:338

Shaw DM (1970) Trace element fractionation during anatexis. Geochim Cosmochim Acta 34:237 – 243

Shaw DM, Dostal J, Keays RR (1976) Additional estimates of continental surface Precambrian shield composition in Canada. Geochim Cosmochim Acta 40:73 – 83

Silver LT (1980) Problems of pre-Mesozoic continental evolution. In: Continental tectonics. National Academy of Sciences, Washington, DC, pp 26 – 29

Silver LT, Deutsch S (1963) Uranium-lead isotopic variations in zircons: a case study. J Geol 71:721 – 758

Smith CB (1983) Pb, Sr, and Nd isotopic evidence for sources of southern African Cretaceous kimberlites. Nature 304:51 – 54

Sobotovich EV, Kamenev Ye N, Komaristyy AA, Rudnik VA (1976) The oldest rocks of Antarctica (Enderby Land). Int Geol Rev 18:371 – 388

Spedding FH, Voight AF, Gladrow EM, Sleight NR (1947) The separation of the rare earths by ion exchange. II. Neodymium and praseodymium. J Am Chem Soc 69:2786–2792

Spohn T, Schubert G (1982) Modes of mantle convection and removal of heat from the earth's interior. J Geophys Res 87:4682

Staudacher T, Allegre CJ (1982) Terrestrial xenology. Earth Planet Sci Lett 60:389–406

Staudigel H, Hart SR, Richardson SH (1981) Alteration of the oceanic crust: processes and timing. Earth Planet Sci Lett 52:311–327

Staudigel H, Zindler A, Hart SR, Leslie T, Chen C-Y, Clague D (1984) The isotope systematics of a juvenile intraplate volcano: Pb, Nd, and Sr isotope ratios of basalts from Loihi seamount, Hawaii. Earth Planet Sci Lett 69:13–29

Steiger RH, DePaolo DJ (1986) Genealogy of the basement of the central Alps. Terra Cognita 6:127

Stewart BW, DePaolo DJ (1987) Nd and Sr isotopic evidence for open system behavior in the Skaergard intrusion. Geol Soc Am Abs 18:764

Stille P, Unruh DM, Tatsumoto M (1983) Pb, Sr, Nd and Hf isotopic evidence of multiple sources for Oahu, Hawaii basalts. Nature 304:25–29

Stille P, Unruh DM, Tatsumoto M (1986) Pb, Sr, Nd and Hf isotopic constraints on the origin of Hawaiian basalts and evidence for a unique mantle source. Geochim Cosmochim Acta 50:2303–2319

Stolper EM (1980) A phase diagram for mid-ocean ridge basalts: preliminary results and implications for petrogenesis. Contrib Mineral Petrol 74:13–27

Stolper EM, Walker D, Hager BA, Hays JF (1981) Melt segregation from partially molten source regions; the importance of melt density and source region size. J Geophys Res 86:6261–6271

Stosch HG, Carlson RW, Lugmair GW (1980) Episodic mantle differentiation: Nd and Sr isotopic evidence. Earth Planet Sci Lett 47:263–271

Sun SS, Hanson GN (1975a) Evolution of the mantle: geochemical evidence from alkali basalt. Geology (Boulder) 3:297–302

Sun SS, Hanson GN (1975b) Origin of Ross Island basanitoids and limitations upon the heterogeneity of mantle sources for alkali basalts and nephelinites. Contrib Mineral Petrol 52:77–106

Taylor HP (1980) The effects of assimilation of country rocks by magmas on the $^{18}O/^{16}O$ and $^{87}Sr/^{86}Sr$ systematics in igneous rocks. Earth Planet Sci Lett 47:243–254

Taylor HP, Gianetti B, Turi B (1979) Oxygen isotope geochemistry of the potassic igneous rocks from the Roccamonfina volcano, Roman comagmatic region, Italy. Earth Planet Sci Lett 46:81–106

Taylor PN (1975) An early Precambrian age for migmatitic gneisses from Vikan: Vesteralen, North Norway. Earth Planet Sci Lett 27:35–42

Taylor SR (1964) Abundances of chemical elements in the continental crust. Geochim Cosmochim Acta 28:1273–1285

Taylor SR (1975) Lunar science: a post-Apollo view. Pergamon, New York, 372 p

Taylor SR (1977) Island arc models and the composition of the continental crust. In: Talwani M, Pitman WC (eds) Island arcs, deep sea trenches and back-arc basins. American Geophysical Union, Washington, pp 325–336

Taylor SR (1978) Geochemical constraints on melting and differentiation of the moon. 9th Proc Lunar Planet Sci Conf, pp 15–23

Taylor SR (1982) Planetary science: a lunar perspective. Lunar and Planetary Institute, Houston, 481 p

Taylor SR, White AJR (1965) Geochemistry of andesites and the growth of continents. Nature 205:271–273

Taylor SR, McLennan SM, McCulloch MT (1983) Geochemistry of loess, continental crustal composition and crustal model ages. Geochim Cosmochim Acta 47:1897–1906

Tera F, Wasserburg GJ (1974) U-Th-Pb systematics on lunar rocks and inferences about lunar evolution and the age of the moon. 5th Proc Lunar Planet Sci Conf, pp 1571–1599

Tera F, Papanastassiou DA, Wasserburg GJ (1974) Isotopic evidence for a terminal lunar catalysm. Earth Planet Sci Lett 22:1–21

Torgerson T, Lupton JE, Sheppard D, Giggenbach W (1982) Helium isotope variations in the thermal areas of New Zealand. J Volcan Geotherm Res 12:283 – 298

Turner FJ, Verhoogen J (1960) Igneous and metamorphic petrology, 2nd edn. McGraw-Hill, New York

Turner G (1970) Argon-40/Argon-39 dating of lunar rock samples. Proc Apollo 11 Lunar Sci Conf, pp 1665 – 1684

Viljoen MJ, Viljoen RP (1969) The geology and geochemistry of the lower ultramafic unit of the Onverwacht Group and a proposed new class of igneous rock. Geol Soc S Afr Spec Publ 2:55 – 85

Vollmer R (1976) Rb-Sr and U-Th-Pb systematics of alkaline rocks: the alkaline rocks from Italy. Geochim Cosmochim Acta 40:283 – 295

von Drach V, Marsh BD, Wasserburg GJ (1986) Nd and Sr isotopes in the Aleutians: Multicomponent parenthood of island-arc magmas. Contrib Mineral Petrol, 92:13 – 34

Wager LR, Brown GM (1967) Layered igneous rocks. Freeman, San Francisco, 588 p

Walker D, Longhi J, Hays J (1975) Differentiation of a very thick magma body and implications for the source regions of mare basalts. Proc Lunar Planet Sci Conf 6th, pp 1103 – 1120

Warren PH, Wasson JT (1979) Effects of pressure on the crystallization of a "chondritic" magma ocean and the implications for the bulk composition of the moon. Proc Lunar Planet Sci Conf 10th, pp 2051 – 2083

Wasserburg GJ (1966) Geochronology and isotopic data bearing on the development of the continental crust. In: Advances in earth sciences. MIT, Cambridge, pp 431 – 459

Wasserburg GJ, DePaolo DJ (1979) Models of earth structure inferred from neodymium and strontium isotopic abundances. Proc Natl Acad Sci USA 76:3594 – 3598

Wasserburg GJ, Hayden RJ (1955) Age of meteorites by the Ar40-K40 method. Phys Rev 97:86 – 87

Wasserburg GJ, Papanastassiou DA, Nenow EV, Bauman CA (1969) A programmable magnetic field mass spectrometer with on-line data processing. Rev Sci Instrum 40:288 – 295

Wasserburg GJ, Jacobsen SB, DePaolo DJ, McCulloch MT, Wen T (1981) Precise determination of Sm/Nd ratios, Sm and Nd isotopic abundances in standard solutions. Geochim Cosmochim Acta 45:2311 – 2323

Weill DF, McKay GA (1975) The partitioning of Mg, Fe, Sr, Ce, Sm, Eu, and Yb in lunar igneous systems and a possible origin of KREEP by equilibrium partial melting. Proc Lunar Planet Sci Conf 6th, pp 1143 – 1158

Wetherill GW (1976) The role of large bodies in the formation of the earth and moon. Proc Lunar Planet Sci Conf 7th, pp 3245 – 3257

Wetherill GW, Aldrich LT, Davis GL (1955) Ar-40/K-40 ratios of feldspars and micas from the same rock. Geochim Cosmochim Acta 8:171 – 172

White WM (1985) Sources of oceanic basalts: radiogenic isotopic evidence. Geology (Boulder) 13:115 – 118

White WM, Dupre B (1986) Sediment subduction and magma genesis in the lesser Antilles: isotopic and trace element constraints. J Geophys Res 91:5927 – 5941

White WM, Hofmann AW (1979) Geochemistry of the Galapagos Islands: implications for mantle dynamics and evolution. Ann Report of the Director, Department of Terrestrial Magnetism, Carnegie Institution of Washington, pp 596 – 606

White WM, Hofmann AW (1982) Sr and Nd isotope geochemistry of oceanic basalts and mantle evolution. Nature 296:821 – 825

White WM, Patchett PJ (1984) Hf-Nd-Sr and incompatible-element abundances in island arcs: implications for magma origins and crust-mantle evolution. Earth Planet Sci Lett 67:167 – 185

Whitford DJ, White WM, Jezek PA (1981) Neodymium isotopic composition of Quaternary island arc lavas from Indonesia. Geochim Cosmochim Acta 45:989 – 995

Wilshire HG (1984) Mantle metasomatism: the REE story. Geology (Boulder) 12:395 – 398

Wilson HW, Daly NR (1963) Mass spectrometry of solids. J Sci Instrum 40:273 – 288

Wilson JT (1963) Evidence from islands on the spreading of ocean floors. Science 197:536 – 538

Windley BF (ed) (1976) The early history of the earth. Wiley, New York, 619 p

Wise DU (1974) Continental margins, freeboard and the volumes of continents and oceans through time. In: Burke CA, Drake CL (eds) The geology of continental margins. Springer, Berlin Heidelberg New York, 1009 pp

Wright PM, Steinberg EP, Glendenin LE (1961) Half-life of samarium-147. Phys Rev 123:205−208

Wyllie PJ (1979) Petrogenesis and the physics of the earth. In: Yoder HS Jr (ed) The evolution of the igneous rocks. Princeton University Press, Princeton, New Jersey, pp 483−520

Xuan H, DePaolo DJ (1984) Isotopic and chemical study of Paleozoic and Mesozoic granitoids, Fujian Province, Peoples Republic of China. EOS 65:1151

Xuan H, Ziwei Bi, DePaolo DJ (1986) Sm-Nd isotope study of early Archean rocks, Qianan Province, China. Geochim Cosmochim Acta 50:625−631

Zartman RE (1964) A geochronologic study of the Lone Grove pluton from the Llano uplift, Texas. J Petrol 5:359−408

Zartman RE, Wasserburg GJ (1969) The isotopic composition of lead in potassium feldspars from some 1.0 b.y. old North American igneous rocks. Geochim Cosmochim Acta 33:901−942

Zindler A, Hart SR, Brooks C (1981) The Shabogamo intrusive suite, Labrador: Sr and Nd isotopic evidence for contaminated mafic magmas in the Proterozoic. Earth Planet Sci Lett 54:217−235

Zindler A (1982) Nd and Sr isotopic studies of komatiites and related rocks. In: Arndt NT, Nisbet E (eds) Komatiites. Allen & Unwin, Boston, 399 p

Zindler A, Hart S (1986) Chemical geodynamics. Ann Rev Earth Planet Sci 14:493−571

Zindler A, Jagoutz E (1988) Mantle cryptology. Geochim Cosmochim Acta 52:319−333

Zindler A, Hart SR, Frey FA, Jakobsson SP (1979) Nd and Sr isotope ratios and rare earth element abundances in Reykjanes Peninsula basalts: evidence for mantle heterogeneity beneath Iceland. Earth Planet Sci Lett 45:249−262

Zindler A, Jagoutz E, Goldstein ST (1982) Nd, Sr and Pb isotopic systematics of a three component mantle: a new perspective. Nature 298:519−523

Zindler A, Staudigel H, Hart SR, Endres R, Goldstein S (1983) Nd and Sr isotopic study of a mafic layer from the Ronda ultramafic complex. Nature 304:226−230

Zindler A, Staudigel H, Batiza R (1984) Isotope and trace element geochemistry of young Pacific seamounts: implications for the scale of upper mantle heterogeneity. Earth Planet Sci Lett 70:175−195

Subject Index